Acknowledgments

I want to thank the many people who helped me with my book, especially my wife, Vera, for being patient with me, in my many hours on the computer and for reading and correcting my mistakes and as I went exploring in Ohio when we visited her family, I especially want to thank my nieces Velma Peck and Emma Hochstetler for doing the excellent illustrations and paintings, and to Rachel for making the last chapter a wonderful ending, and for the many Ohio Amish friends and 'frienshaft' that I met along the way and their helpfulness in my researching family history. Also my nieces and nephews who contributed stories. The Indiana Amish friends and relatives and especially the Amish Libraries and the staff who were very helpful with my research. I want to dedicate this book to the future generations, who ask where do we come from and who are we? I also dedicate this to my family and the future, my son, my grand-daughter and my two great-grand-daughters so they will know where we came from. And help guide them to where they are going.

Uncle Chris

IV

A Journey to the Future

Chris J. Miller

Best wishes

FROM THE EARLY CHURCH TO THE PRESENT TIME,
ONE FAMILY'S STORIES OF FAITH AND HOPE FOR THE FUTURE

© 2017 Chris J Miller

All rights reserved. No part of this publication may be reproduced, stored in a retrieval system or transmitted, in any form, or by any means, electronic, mechanical, photocopying, recording, or otherwise, without the prior permission of the publishers.

For more information contact:
Chris J Miller
1549 Redbud CT
Goshen, IN 46526

Illustrations & Paintings: Velma Peck & Emma Hochstetler

Table of Contents

Preface VII

The Early Church 1

The Anabaptist 7

Coming to America 19

Samuel (Mueller) Miller 27

Christian (Schmidt) Miller 33

Jonathan Miller 41

Christian (Christal) J. Miller 47

Christian C Miller "Iowa Christ" 51

"Broad Run" John 59

Bishop Levi "Leff" Miller 65

Bishop "Leff John" Miller 85

John J.L. and Mary Miller 93

The Christner Family Branch 99

Jacob J.C. Miller (Jake) 109

Chris Jay Miller Story 125

Robert Dean Miller 143

Rachel Miller-Minick 147

Many Miller Memories 153

Epilogue 201

Preface

One of my nieces asked me one day, "Uncle Chris, can you tell us about our family history? About where they came from or anything about them or stories of our ancestors? And what were they like?" Well that got me to thinking, I used to ask my older brothers or sisters if I wanted to find anything about my relatives, then I realized there is no one left to ask. Then I decided to start writing things down before I forget also. My Dad was a great story teller and he always had a ready story wherever he was at, be it with family or with friends at the local filling station, unfortunately I don't remember lots of the stories, only bits and pieces so I have to fill in the spaces best as I can.

Mom told us many stories of her childhood and of her ancestors along with words of wisdom and advice about life. Mom told us some stories that we took to be old wives tales, but surprising most of them turned out to be true (baby Jacob Christner, and the man buried outside the graveyard fence, etc.) by researching and many trips to Holmes County, Ohio, and with the co-operation of many people, (especially the Amish historians) I came up with many interesting stories.

I tried to come up with as many interesting stories about our ancestors as I could rather than genealogy as I think this puts a little life into our ancestry.

The Bible has many stories about ordinary people in their everyday life, Jesus told parables about every day people and happenings. I believe the stories did not stop with the Bible. I think the stories continue even today and always will. Everyone has a story maybe good or bad, never the less a story. There are many people of the Bible that I call heroes, people that left an impression, good or bad.

Through-out history we can read of many such people, some who stood up for their belief and paid the ultimate price, yet would not waver, even at the last minute. There are some who dedicated their lives for service to others and some who preached and tried to reach as many unsaved souls as they could in there lifetime. And there were those who were of the vilest nature trying to destroy as many people and souls as they could, they must surely have been influenced by Satan the ultimate destroyer of mankind and the enemy of God and must now pay the price.

I tried to find stories of my ancestors that I could, I was handicapped because until the present time, every day history was mostly handed down orally from generation to generation. Therefore many stories were lost to the graves. Especially at the time when our ancestors migrated to America, they had to leave at a short notice and could only take what was essential and when they landed they had to establish a new home in the new country and that left little time for writing, also it was and still is much easier to talk than to put things down in writing.

In researching I tried to recall as many stories as I could remember from my parents and older siblings and I asked my nieces and nephews for stories that they remembered, some stories I gathered from family books and other sources. When we traveled to Ohio to visit family I would take time and interviewed Amish people there, they were very kind and helpful with information that they could, and I got a lot of information but sometimes I could not get the stories about some of the people I wanted. I was told I would have to ask some old people who would remember stories of long ago, only thing is I couldn't find many older people!

By researching family books, many of which have small stories tucked among the names, dates and relatives, also many of the newer family books devote a lot of space which is very helpful in finding stories first hand. In the meantime I also found a number of other interesting stories not related to my family which means more books could be written some other time.

My hope is that the younger generations and generations to come will be able to understand where their ancestors came from and what they were like. Many of our personalities, traits and habits, family facial resemblance and body stature, DNA signature are all connected to our ancestors as well as to our siblings, cousins, and children.

Like it or not we have relatives! Unlike friends which come and go, relatives are there to stay no matter what, they will always be relatives. Sometimes we can't get along and sometimes we're best of friends. But we're in this together we might as well make the best of it.

I believe our ancestors were concerned and prayed about not only their children but also their future generations. The Anabaptists chose to live the way they believed and showed us how to die for their faith, without wavering as the flames curled about their feet, they sang songs of faith and encouraged their peers as God took away the hurt of the fire, to the amazement of the authorities the number of faithful grew by leaps and bounds. Later as they decided that this would not be the place to live and raise their children and made a decision to move to America leaving the home country and burial grounds of their ancestors never to return again leaving behind some of their family and friends to travel many miles and cross a great ocean to seek a new and safe life for themselves and the future generations. Our Amish ancestors were known to always include a prayer for their children and generations to come. It should be comforting to think that someone was concerned about us before we were even born.

I for myself am very glad to be a part of a family of God!

Uncle Chris

BOOK ONE

The Early Church

Mom's Bibel

The Bible has many stories about ordinary people, and how they interacted with God, many stories of hero's from children's Bible stories to stories of people who experienced God's wrath when they would not obey Him, But great stories they are, with real live heroes that outdo modern comic heroes, like Goliath, Samson and Jesus. The stories did not end with the Bible, but they continue on to the present day and beyond into the future, we can look to the book of Revelations and see what the future holds. I find it very interesting to search for stories and hero's, down through the ages and some of them could have been or were some of our very ancestors. To find just a little glimpse into the lives of our ancestors brings us just a little closer and makes history a little more real and a whole lot closer. The Jewish people always valued their genealogy and could trace their linage all the way from Adam and Eve to Jesus Christ, although we can not trace our genealogy as well, we should be just as interested who our forefathers were. If we don't know where we came from how shall we know which way to go ? History is such a good teacher, we can learn from the past so we don't make the same mistakes in the future, Reading about the past can help guide us thru life.

There are many-many stories from the Biblical times to present, and I will only attempt to bring you a few of them, especially the ones that would relate to our family history

The first book of the Bible begins with the Majestic words " In the beginning God created Heaven and Earth" The book of Revelation, the last book of the Bible, now had the reassuring ending "The Grace of our Lord Jesus Christ be with you all, Amen." What a great story and blessed assurance we have to live by.

The last book in the Bible prophesies the final defeat and banishment of evil, and the final everlasting triumph of truth, love and righteousness, Evil, while winning many battles, is still fighting a losing war. The battle with evil will continue until history ends, but the outcome is already decided. The Church built by Christ will eventually triumph; "the gates of hell shall not prevail against it" (Matt.16-18).

The spread of Christianity had it's origin in a tragedy—the cross. The world's greatest tragedy on a Friday was changed to the greatest miracle since creation, only several days later. Christianity is a belief in resurrection. You cannot be a Christian and not believe in immortality.

Jesus' mission to the earth was to teach love and kindness, to set an example for humility, and to give his life as a ransom for all mankind under the slavery of sin, which is a sickness of the soul and the spirit.

Several weeks after Jesus rose from the dead, at the time of the Jewish festival of Pentecost, the Holy Spirit came upon Christ's followers, as he had promised it to them. We read about God's Spirit on people in the Old Testament times, but the first Holy Spirit on a whole congregation was at Pentecost. The Holy Spirit came not only to the twelve disciples, as some believe, but to all who were assembled in the Upper Room (acts.2:4) just as it is possible in a worship service today!!!

After the appearance of the Holy Spirit upon the disciples, they were changed and recharged. No longer did they fear the persecution and suffering they knew might be in store for them. And the day of Pentecost did not end when the day was over.

The Day of Pentecost with the Holy Spirit coming upon Jesus' followers to comfort and guide them, and help them resist the temptations around them, will continue until the Judgment Day. That will be eternal, and have no evening following it.

In Christianity is the joy of being at peace with God and fellow believers, Here is a peace of mind with a joyful trust in a Savior. Here is a feeling of security in a cruel, sensual and evil world. Christians have no fear of the future or of death. There is no mention of missionaries in the early Church, Christianity was spread by ordinary daily socializing. People liked to be around "those people who were friends of Christ" More effective than the full time missionary for the spread of Christianity was the attraction of the day by day living witness of the common people. Christianity is it's own best missionary. Most of the Christians were just ordinary working people whose upright, loving and sincere living stood out and could not help but be noticed. Many thousands upon thousands of people over the centuries were to learn that one of the courses taught in the school of Christianity was suffering and persecution. This could be any thing from reproach and scorn to suffering, torture and a martyr's death.

The first persecution of the Christian church started just shortly after Jesus ascended to Heaven. After the birth of the Christian church at Pentecost, the apostles were jailed and beaten by the Jewish priests and religious authorities, and were commanded to cease speaking about the name of Jesus.

After the death and resurrection of Jesus Christ, his followers called for all to turn from sin, and accept God's gift of salvation. Baptism into this persecuted church was a voluntary choice to live the way of Christ and identify with a new community of believers. Despite severe persecution Christianity spread rapidly in the Mediterranean area and beyond.

The first martyr of the Christian church was one of it's deacons, called Stephen. His enemies " were not able to resist the wisdom and the spirit by which he spake" (Acts 6:10). Persecution by the Jews and the Roman Empire continued for nearly three hundred years.

In the year 313 Roman Emperor Constantine issued an edict of toleration granting freedom to the Christian faith, By the end of the century Christianity had become the religion of the Roman Empire, as the state and church united, baptism ceased to be a voluntary choice. Now it was expected for all citizens at birth. In essence every Roman citizen became a Christian, every Christian a Roman Citizen. Now the Emperor Constantine thought of himself as the head of the Christian Church which was now in partner-ship with the Roman Empire. Children who were baptized as infants had no political or spiritual worries, nor choices.

Constantine

This thousand year reign lasted until the invention of Gutenberg's movable printing press in1455, the people searched in vain but could not find passages saying they should be baptized at birth, but it said baptism should be free and personal and not dictated by the church.

Martin Luther

When Martin Luther nailed his ninety-five theses on the door of the church at Wittenberg, Luther, the former priest now had to go into exile, when he did he started translating the Bible into German. He was probably one of the few people at that time that understood the four languages, Hebrew, Greek and Latin and was able to write it in German which was the common language of the day. Amazing the German Bible is still being used today by many of the Amish and the Bible I got from my parents is called; Die Bibel — Heilige Schrift by Dr. Martin Luthers. Printed in 1907.

Some people were being convinced that the Catholic Priests were not teaching the Bible correctly, At that time they were using the Latin version of the Bible, and the common people were not allowed read from the Bible so therefore the Pope and Priests dictated what was taught to the people. Only two years after the first printed Bibles were available to the common people, the people realized that some of the things they were being taught was not what the Bible said at all. One of the things was infant baptism, no-where could they find any scripture to that effect, but they did find scripture about faith baptism by responsible adults.

Mom's Bibel page

Some people started holding secret meetings and Bible studies and were soon convinced they should be properly baptized or re-baptized. Knowing they would be severely punished for disobeying the church, they felt convicted that this they must do, to follow the word of God rather than the preaching of man.

This then was the beginning of the Anabaptist movement, the breaking away from the Catholic Church and the Roman Empire. Most churches of today can relate back to their Anabaptist background.

You are not ready to die until you are ready to live,
NEITHER ARE YOU READY TO LIVE
until you are ready to die.
Clarence Yoder

You cannot neglect to read the Bible and
REAP A WELL-GUIDED LIFE.

It's always good to be careful
when you give advice,
because someone may take it.

BOOK TWO

The Anabaptists

On a Saturday evening of January 25, 1525, in the depths of the Swiss Alps a handful of people gathered in the home of Felix Mantz where they had came together to study the Bible, and nowhere could they find anything concerning the baptism of infants. They came to a conclusion that a Christian church was to be composed only of those people who were adult enough to see that they want a change in their life and ask to be baptized in the name of the Father, Son and Holy Spirit. The pattern was to believe first then, be baptized.

A great anxiety came upon them all. They fell upon their knees in prayer and asked God to fulfill His will and divine guidance. After they had risen Georg Blaurock, the former Catholic priest asked Conrad Grebel to baptize Him which he did. A pitcher of water and a cup were brought and with holy awe Conrad Gerbel baptized

Georg Blaurock

Georg Blaurock, Then Georg baptized Conrad Grebel and Felix Mantz, Then at least 15 others were baptized also.

The new brothers and sisters in Christ expected persecution in some form or other. And come it did. Almost immediately the small group was put to a fierce and bloody persecution that but for God's providence on them would have exterminated the cause of adult baptism. Even though the persecutions were harsh and severe this did not slow down growth of the Anabaptist in Europe, in fact every time they tortured someone in public many of the onlookers became convinced that this was not right, and many times there were those who decided to join the Anabaptists then and there, even the executioners many times converted and were later persecuted themselves. As they were tortured many would either sing hymns or encourage others so much that they would either put a clamp on their tongues or remove it to keep them quite. Many times they showed signs of having no pain, as one story goes, a man chained to the stake and flames dancing all about him reached into the flames like he was washing his hands in the flame apparently without any pain.

The Martyrs burned at the stake

MICHAEL SATTLER

Michael Sattler was one of the Martyrs that you don't often hear about, but he was one who helped organize and steer the Anabaptist in the right direction almost more so than many of the others. Though he only lived thirty-two years and died a horrible death at he hands of the executioners but his faith never wavered.

Michael was born about 1495, became a respected and a learned scholar. He studied the scriptures and read the writings of the early church fathers.

Wanting to live a deeper Christian life, he decided to enter a Catholic monastery. He soon elevated to office of prior, Michael soon learned that the life in monastery was not as he thought it would be. Many Monks had mistresses and lived lives of drunkenness, and sinners. Michael could hardly believe all that he saw and heard.

In 1523 Michael left the monastery, along with the comfortable life and promising career before him. He married Margaretha a former nun, They wanted to be with people who took the living of Christianity seriously, in 1525 they joined the Anabaptist, who had been organized only a few months earlier. Michael soon became a spokesman for the Anabaptists. The many people coming from such a variety of backgrounds, they had to have direction.

A number of Anabaptists met at Schleitheim on the Swiss-German border to discuss their differences. Michael, wrote out a simple statement of faith of seven articles that are based on the Bible. This "confession of faith" was approved as written by everyone present. It was soon printed and widely distributed among friends and enemies alike.

This "Brotherly Agreement" or the Schleitheim Confession of Faith as it is called. Was more like a letter to friends of like faith than like a written church standard or a formal declaration of faith. (it is still being printed and used today). The printed and widely distributed copies of the pamphlet were soon to be read wherever the scattered Anabaptist congregations happened to be. A few copies fell into the hands of the state-church authorities who were alarmed at the stream of people deserting their churches to join the Anabaptists. To the Catholics all seven articles were the

Bruderlich Vereinigung

THE ANABAPTISTS

height of heresy. Less then a month later on his way home from the Schleitheim meeting Michael was betrayed to the authorities.

He and his wife were arrested with a group of other Anabaptists. A complete copy of the Schleitheim Articles was found on his person.

As Michael was known to be the writer of this "heretical" booklet which was thought to be misleading many people, he and his wife were kept in a high security prison. Despite receiving a large number of appeals for mercy for them, all this did not reverse the decision they had already made about the despised and hated Anabaptists.

Under a guard of 24 armed men the trial of Michael and his wife, along with nine other men and eight women, began. Michael was especially singled out as a leader of the despised Anabaptists. In the daily malice against him Michael was shouted at, threatened and mocked by the hostile court, but his complete self-control only aroused the judges to even more hate.

Throughout all this, Michael's calmness and composure never left him. He did not have a trace of fear for what he knew what was ahead for him. He could not be provoked to any hateful words or actions, and he seemingly knew no fear.

After the charges were read and Michael had answered them. The judges left the room to decide the fate of Michael Sattler and the others with him. They returned with the verdict an hour and a half later.

With all in heavy silence the gruesome sentence was read *"Between the representatives of his Imperial Majesty and Michael Sattler, judgment is passed that Michael Sattler shall be delivered to the executioner who shall firstly cut out his tongue; then throw him upon a cart and with red hot tongs tear pieces out of his body twice, and on the way to the place of execution make use of the tongs five times more in like manner. Thereupon he shall burn his body to a powder as an arch heretic."*

Sentence was also passed on to his wife and the other Anabaptists with him, the men were executed with the sword and the women by drowning. Michael told his friends that if he was able to, he would raise two of his fingers heavenward in the flames to let them know that such a death was bearable and that he still kept his faith. When the ropes with which his hands were bound, were burned through, he used his last remaining strength to raise one arm upward with two fingers spread out as he had said he would do.

Michael's torture and death were meant by the authorities to frighten others from joining the rapidly growing Anabaptist groups. However, a seed was now sown, the letter of encouragement which Michael had written to the brotherhood, along with an account of his unjust trial and heroic death, was soon printed and widely distributed and read. This small booklet of 36 pages was to stir up a backlash against the state churches by a public that became repulsed at the cruel and ferocious methods of keeping the people of the churches under the thumb of their high church authorities. Many people became sympathetic to the Anabaptists and left the state churches. They joined the Anabaptists because of this lost confidence in their religious leaders who abused their authority because they were afraid of losing power.

To the dismay of the state-church authorities, the Anabaptist faith spread among the common people with surprising rapidity. They noticed the complete absence of fear, even when they faced such an ordeal as they had seen or read of what Michael Sattler went through.

No, the deaths of Michael Sattler, his wife and their friends were not in vain. It bore fruit in the church groups of today that are descended from the Anabaptists.

Michael Sattler in his short life left a legacy that we can live by even to this day of uncertainty, when Christians are being persecuted thru-out the world, we need to realize that we may have to pay the ultimate price. Would we have strength and faith when the flame came about your toes? The Devil can get you where-ever you are and he is lurking right there at the last minute ready to grab you.

Not only were they persecuted but their wealth and property was confiscated and their children were taken away and placed in other homes by authorities. Some Anabaptists were escorted to the border and were warned never to come back* *In 1660 a Christian Christner was taken prisoner at Bern, Switzerland, he was expelled from his homeland and placed aboard a boat on the Aare and Rhine Rivers.*

When the authorities decided that public persecution was not the answer, they decided to put about 50 some people, men and women some as young as 16–18 yrs. old into a dungeon at the Passau castle, and gave them no future, no time of release and no guarantee if they would come out alive. But what these people did was to start singing to while the time away, knowing that the guards in the upper floors could hear them, and they sang in such a way the guards could not understand them, all they knew it was beautiful singing from the dark dungeons below. After about five years they were released and they wrote and assembled those songs of faith and hope into a song book called the Ausbund which the Amish still use as their song books today as well as the unique way of singing as they did back in the dark days in the Passau castle. It is the oldest known song book still being used today. What a unique story to pass on to the future generations! Today when these songs are sung in the Amish church there is beauty in voices blending together (men and women)

1736 Ausbund

singing a familiar song in harmony without interference of musical instruments and remembering that these songs were sung by their own ancestors many generations ago while their very lives were in uncertain future. We hope they never forget the legacy of their forefathers.

Today on any given Sunday, about 1,100 Amish churches will be singing the same song at the same time, the next Sunday the other churches will be singing the same song! They always sing the same opening song (O Gott Vater) so if they all start at the same time, God will have his own chorus every Sunday!

Another legacy our plain German-speaking owe to Michael Sattler is his love for songs and singing. He is known to have written at least one of the hymns in our Ausbund hymn book, the one on page 46 which starts out as; *"Als Christus mit seiner wahren Lehr." (When Christ with his true teaching came)*

How touching then would have been a parting in Michael's time when no one would know who would be the next martyr? Only God knows how many silent tears were shed for each other while they sang a parting hymn together. In Michael's parting hymn the complete acceptance of God's will shines through. Even in their uncertain times of risk and peril they sang of the holy joy which can, by God's grace, lead to eternal rapture.

As the movement of re-baptism advanced more priests left the Catholic church, as well as other citizens and joined with the Anabaptist, and left their mark in history. There was Martin Luther, Ulrich Zwingli, Michael Sattler, Menno Simons, Peter Waldo, John Huss, John Wycliffe, Jacob Hutter and Jakob Ammann and others. It seems that most of these learned men had their own ideas and their interpretation of the Bible.

In 1693 there was a formation of two Mennonite groups when Jacob Ammann, a minister, was concerned that not all of the articles of the Dordrecht confession of faith, (written in 1632) were being put into practice. He, along with a number of other more conservative leaders, were interested in and sought for real renewal in the brotherhood, For them the real issue was whether we are going to keep the way of holiness, The Swiss Brethren had not subscribed to this confession like the other European Mennonites had. The greatest item of disagreement grew out of Ammann's conviction of the "Meidung" or shunning. (Matt. 18: 15-18 & 1 Cor 5:11). This was a method of avoiding one who had been excommunicated because of departing from the faith.

Jacob Ammann also believed in plain clothes, (no buttons), foot washing, avoiding contracts with outsiders, holding church services in homes rather than in elaborate sanctuaries. Practices which continue to this day in the Amish church. This polarization resulted in two factions, the Amish- Mennonites and the Mennonite church.

As Jacob Ammann and Michael Sattler's writings guided the Amish people, Menno Simons and Conrad Grebel were organizing the Mennonites, although they did not agree on certain principals and sometimes avoided each other, they have over the years lived in close harmony and respected each others opinion.

In the migration to America they mostly settled close their Amish brothers, with the Mennonites being more liberal, settled in the urban areas and blended more in the American culture, their migration pattern across the U.S. was very similar to the Amish, especially those in farming and agriculture. Another interesting fact is that mostly when Amish wanted to leave the Amish for various reasons that usually joined one of the nearby Mennonite churches, which were more liberal with similar spiritual values. There were several attempts made by the Amish-Mennonite leaders to reconciliation with no avail.

Catherine Muller - Martyrs Mirror

PERSECUTIONS OF CHRISTIANS IN PALATINE AND SWITZERLAND

(There is little documentation from that period of time. There is only one source of Chronicles (that I know of) that was not lost or destroyed. Frank Eshelman lived in Switzerland in those hardship years and he kept record of what was going on in Europe. Those stories were assembled and published in

The Budget many years ago.)

In Dec. 1644 a new law went into effect, forbidding any future marriages to be performed in their old Mennonite church except by government authorities and clergymen.

The law demanded that all the Anabaptists were compelled to make list of every one of their children who were not baptized as infants. Also of all marriages among them that were conducted by their own Mennonite Bishops, were declared illegal, they were now ordered to be canceled and the children of such families were pronounced illegitimate . the parents were ordered to be separated from each other and put in prison. Their children were ordered to be taken away from them and put into the Orphan Asylum, where they were to be baptized and educated in the Romish church customs.

All the Bishops, preachers or church leaders who had performed such marriages shall be branded with red hot irons and cast into prison. The principal object of this law was to prevent farther development of the Swiss Mennonite church in Europe.

This unmerciful law brought still more great fear and sorrow into the homes and hearts of thousands of innocent Christians, who had already suffered many severe punishments.

The 30 years of war, beginning in 1622 the greatest destruction was begun, huge armies led by Tilly brought the land to a homeless waste. Ten years later the Spanish Galles again brought more destruction on the land, which left but the remains of glowing iron and the ruins of stone dwellings. In 1639 the French and Bavarians ravaged the land again and nearly all the crops were destroyed which brought more great famines for lack of food. In 1644 to 1645 nearly all of the land lay in waste like a great desert, neither friends or foes cared to establish any homes here. Germany and Switzerland lay as a land of ruins. Most of the people and lots of the property was gone.

Many of the suffering refuges were living in terrible hardships in the Palatines adjoining Holland. In 1674 Turene with a large army swarmed through the one side of Rhine Valley and made a clean sweep and complete destruction then they crossed the Rhine and ravaged the other side of the Rhine with similar destruction. Those who were fortunate enough to escape the swords and spears were hiding in the cliffs and caves in the mountains with almost nothing to eat. The Romish Commanders sent messages throughout the country that in 3 days will be given for everyone to leave the land or be killed.

The remaining suffering refuges now fled in the direction of Manheim, the roads, the countrysides were filled with homeless families. It was just a few days before Christmas and the bitter cold days seemed terrible for the little suffering children who were among them. Thousands of them perished along the way. Men, women and children lay dead in the snow banks everywhere. Many of the fathers had been taken away from their families and locked up in prison in Zurich. Some of them had been killed in the Martyrdoms, especially the ministers and bishops while their families were scattered in every direction. The terrible destruction did not come to an end until 1697, by this time nearly all the remaining Christians had fled out of Switzerland to the Palatines and were known as the Swiss Refuges, among were large numbers of our Mennonite and Amish ancestors.

When William Penn, the noted Quaker preacher, discovered the pitiful condition of these poor refuges, he at once meditated on a plan to help them to his land in America as Christian Refuges. It was a terrible sight to see the poor suffering people of all ages with old ragged clothes which had been worn for a year and no other clothing to replace. At night many of them slept on the bare ground with not a single cover. Most of them so frightened that they did not have a good night's rest in more than a year.

William Penn was surprised to find these people so well educated in scripture. He soon discovered that they were a true, honest class of people. He soon organized the Frankfort company and started his first ship 'Concord' for America, this was in 1683. He had 13 families on his first ship which started from Manheim and landed at Philadelphia, making the 3,000 mile trip safely.

William Penn wished to settle his land with honest Christian people. Later on, he helped to arrange ship loads of suffering refugees and brought them to

Pennsylvania. It was a great help to the early settlers to establish the principles and foundations of Christian Churches in Pennsylvania under the kindness and Brotherhood spirits of William Penn and his Quaker Brethren.

The Mennonites of Holland also opened their hearts and purses to help their suffering Christian Brethren out of Palatinate and Switzerland. Many times these Christians did not want to leave their homeland because their spouses and their children had been taken away and they had no way of knowing where they were and reluctant to leave them behind. Some of the children that were taken away as youngsters, when older they went searching for their parents, for they remembered their parents as they were forcefully taken away and forced to live with other people. Many families were torn apart and never did find out what happened to the rest of their family. But they labored until the last of Amish and Mennonites were out of Switzerland and Palatine. The Holland Dutch also helped the Christians to pay for and find ship passage to America. For this the Amish and Mennonites should be ever grateful.

I VALUE THE FRIEND WHO FOR ME
finds time on his calendar,

but I cherish the friend
who for me does not consult his calendar.

Robert Brault

BOOK THREE

Coming to America

As the leaders and churches became organized and disorganized the persecution did not stop, some of the Anabaptist fled to neighboring countries, Alsace and Palatine and France for short period of time, then followed the Rhine River down through Holland and onto ships sailing for America. From 1700 to 1850, With an invitation from William Penn, to settle on his vast acreage in Pennsylvania, many of our ancestors fled to America in search of religious freedom and a better place to raise their families. Leaving many of their family and friends for an uncertain life in a foreign land, never to see their loved ones and homeland again.

Going to America was no easy choice because it meant leaving not only your homeland but also many of your loved ones that you knew you would likely never see again. You had to choose who in your family could go and who couldn't go, And also what you could take with you which was very limited. Usually just your clothes as needed, A Bible, an Ausbund and The Martyrs

Mirror, if they had them and not much else. No time to write down family history or hardly time to say goodbye.

If someone in the family had health problems at the last minute choices had to be made, If they couldn't go along they had to stay with relatives or with someone else, and then come later.

Then there was acquiring and paying for passage on a ship. If you had money you could pay for your passage, But if you were poor you had to find someone to pay your fare and then become indentured, ie. You had to work for a certain length of time (years) to pay off your debt, as in case of Wilhelm Bender of the Bender family. And it was a uncertain journey across the sea in a tiny wooden  ship at the mercy of the winds, the weather and what time of year it was, they also had to contend with harsh temperatures. I researched some of the ships that our ancestors used there was "Harle" (Jacob Hochstetler), The ship "Phoenix (Christian Miller family), "The Bark Elizabeth" (the Bender family) These are just a few of the ships used by our ancestors. lasting from 30 days to several months until they reached their destination. Then there was sickness and disease from living in close quarters and unsanitary conditions, Youngsters under one year of age had a slim chance of making the voyage. My Mother often told me a story of just such a case. She said there was a mother who had a small child who was sick and the Capitan came around and said there must be somebody sick on board because "whales" (sharks) had been following the ship for sometime and he said if someone is sick or dead he would have to throw them over board as he was afraid they might attack his ship then they would all be in danger. But the mother tucked the little-one under her clothes and close to her body and was able to keep it hid until it regained health and it was able to finish the journey to America. I always thought this was a fascinating and interesting story. Much later, reading the Christner Genealogy book I found that same story as Mom related it to me, and he was my Great- Grand father Jacob Christner!!

After our ancestors landed in America then their problems were to find a locale and a place to live, some had relatives already living here who could help them get started, at that time much of America was still occupied by Native Indians. The government was making tracts of land available to the settlers and thanks to William Penn who had acquired a large tract of land from the British government and was making it available to the Anabaptists.

The urge to push westward seeking new land was there from the start. When our Miller ancestors settled in Berks County, Pennsylvania, they continued practicing their distinctive customs. Their emphasis on non-conformity to the world was reflected in their mode of dress, refusal to adopt to the latest trends, and avoiding outside contracts. They practiced peace and non-resistance such as was illustrated by the Jacob Hochstetler family (part of our ancestors) who refused to use guns to defend themselves from death and capture by the Indians. They bore each other's burdens co-operatively because they were a close knit community,

In 1810 *John Miller* (b. 1766) was living in Quemahoning Township in the northern part of Somerset County, Pennsylvania, where his older brother *Christian "Schmidt" Miller* and his younger brother Isaac had settled, The urge to move further was there.

Later John moved to Tuscarawas County, Ohio see ("Broad Run" John), A number of Amish families moved to Ohio, then later to Indiana, Michigan then Illinois and Iowa. Some also immigrated to Ontario, Canada, (Christner Family). Later on some Amish came to America via New Orleans then traveled up the Mississippi River to Illinois. At one time Illinois had the largest Amish population in the states. The Amish pushed on west ward following the national trend to inhabit the western states, like Kansas, Oklahoma, Minnesota, and all the way to Oregon. Some set down firm roots and stayed there while others because of difficulties in taming the land or other reasons came back where they started from to the comforts of home and relatives.

Conestoga Wagon

Years ago some of the areas where the Amish settled did not survive because of harsh conditions, poor soil or lack of like minded-ness in church matters. Therefore there is a book called *Amish Settlements That Failed*. With a lot of interesting stories of our ancestors and how they coped with broken dreams, hardship and living when our country was still new and untamed.

Even thou there were only about 500 souls, Amish and Amish-Mennonite that migrated from the Alsace- Palatinate area to America, It's amazing how they have grown and flourished since then. Many have called them an outdated and dying culture. Something I have observed, is that when they migrated to America there were three items that almost every family brought with them, that maybe the key to their unique belief and lifestyle thru the years, there was scarce room in their trunks and bags for clothes and necessities but they made room for, a Bible an Ausbund and a Martyr's Mirror, all in German.

The Amish read and reread these books and lived their life around them. To them the Bible (the whole Bible) was history, words from God how to live and worship, words from Jesus teach love, forgiveness and salvation, words how to live and how to die. The Ausbund has words of inspiration, songs of faith and hope, glorifying, worshiping God, Jesus and the Holy Spirit. Singing in harmony with fellow believers in a closeness and comfort of someone's home with out the interference and noise of musical instruments, just as their ancestors did many years ago when the songs were written, as the Anabaptists were worshiping in secret in homes, barns, caves, where ever they could meet, worship and encourage one another.

The Ausbund is still sung today the same as when it was first written around 1540,

Amazing story of faith and hope still being sung in every Amish church service even today wherever Amish people gather across the United States and else where.

The Martyrs Mirror documents in word and pictures (wood cuts) stories what life can be like to be a Christian, the faith that is required and the protection you have in the midst of fire and persecution. And know that God will be there with you faithfully no matter what happens, He can take away the pains of the fire and give you strength and hope to endure through any hardships.

The Pennsylvania Dutch language plays an important part of their life, it is like a comradeship whenever or wherever Amish or anyone who speaks their language meet there is a sense of compliancy and comfort even if you have never met before.

I experienced this phenomena when I was researching for my book, wherever I met with Amish folks (and I am not dressed Amish) I would speak the Pennsylvania Dutch and immediately I was recognized as one of there brothers, whether it was in Pennsylvania, Ohio, or Iowa. We understood each other in language and heritage.

The Amish continue to immigrate until they can now be found in almost every state of the union. Because of choice of lifestyle, livelihood and unavailable land in rural areas, also some chose because of church issues, such as dress codes, use of modern machinery etc. Many young families move to rural areas in other states with others to form a church in the area. A lot of time the local people won't hardly notice them because usually they are quite neighbors and don't make much fuss and associate with each other. They do make for good neighbors.

They are also very enterprising and self sufficient and can usually find something to do wherever they are, to make a living for themselves and their families.

The Amish continue to grow and expand and today they are one of the fastest growing churches in America, Because of their isolation and lack of national organization it surprises many of the facts. One reason for their growth (even though they do not solicit outsiders) they still double in numbers about every twenty years, one reason is because they have an amazing retention rate of about ninety percent of their children who chose to live a life style of their parents and their ancestors.

Some Mennonites from Alsace and Palatine were invited to Russia by Catherine the Great to the steppes of Ukraine and promised religious freedom and freedom from military service. They flourished and grew wealthy, established their own towns and villages. They established their own schools and churches, Many ran their own businesses, they used improved farming methods and became affluent in many ways for a number of years. But after Catherine was disposed of and new rulers came into power, things changed. Later rulers were not as tolerant as Catherine was.

Then there was a time of trial and tribulation as the Czars took over and tried to force the Mennonites into military service and to give up their property to a communist society, which they would not do. That was a time of much suffering and persecution by vandals released by the government. They started looking elsewhere where they could live peacefully.

During WW II and the Stalin era most of the Mennonites had left or were leaving.

Some fled to Alsace, Palatine and Holland, and then started migrating to America. Some of those from Russia settled on the prairies of Manitoba and Saskatchewan in western Canada. Also some of them settled in several places in South America.

While others from Russia went to central Kansas and settled on the prairies there they brought with them (packed in with the families baggage) some of their Turkey Red winter wheat which they had raised successfully in Russia, Now they planted this wheat (before this the farmers had tried to raise corn and other crops which failed miserably, what with dry hot summers and swarms of

Early thresher

locusts) now the wheat grew in the winter and was harvested before the locusts and dry weather arrived. Looking at the great plains, acres of wheat from the Mexico border north into Canada most of that wheat today can trace it's ancestry to the Turkey Red wheat brought over by the Mennonites from Russia.

Although the Anabaptists Mennonites, Amish, Hutterites and The Church of the Brethern (Dunkards) have infiltrated into American society they have been mostly ignored by historians. Because they mostly live in rural areas, don't involve in politics, sports, news media or are involved in crimes or other high profile activities. They still leave a stable, moral and Christian influence wherever they are and live, sometimes misunderstood and ignored even by those who live near them.

Today, as back in the time of the early church many people like to live near "these peaceful people who are friends of Christ"

GOSSIP IS LIKE MUD ON THE WALL—
you can wipe it off, but it leaves a bad spot.

JAKIE SAYS HE GOT A NICE CAMOUFLAGE SHIRT
for his birthday, he hung it up in his closest
and now he can't find it.

BOOK FOUR

Samuel (Mueller) Miller

From the early church history down to the 1900s, accurate records are hard to come by. In early times records and diaries of everyday life were not kept by ordinary people, like they do now. Information that was kept was by churches, towns / villages and state, and it was recorded in the language of the locale and time in history.

Records that you find are rare and often not complete, ship lists usually did not list women and underage boys. It is a matter of guess work and a lot of cross-checking of dates, getting bits and pieces from different sources and trying to make sense of it all, then there is the problem of names, different spellings and no middle name or other identification. I tried as best I could but I will not guarantee to the accuracy of this information.

Miller Family History* *from (Miller Family History, Eli S. and Marie Miller 1981)*

The Amish-Mennonite Mueller-Miller family originated in Switzerland. It is an occupational name, very common in the Germanic speaking world, where it is spelled Mueller. Those who came to America around 1750 can be traced back to the Thun area southeast of Bern, a beautiful region of hills and small farming hamlets, emphasizing dairying and hay crops. The Mueller families joined the Anabaptists in the area along the east bank of the Aare River, near Munsingen and Oberdiessbach.

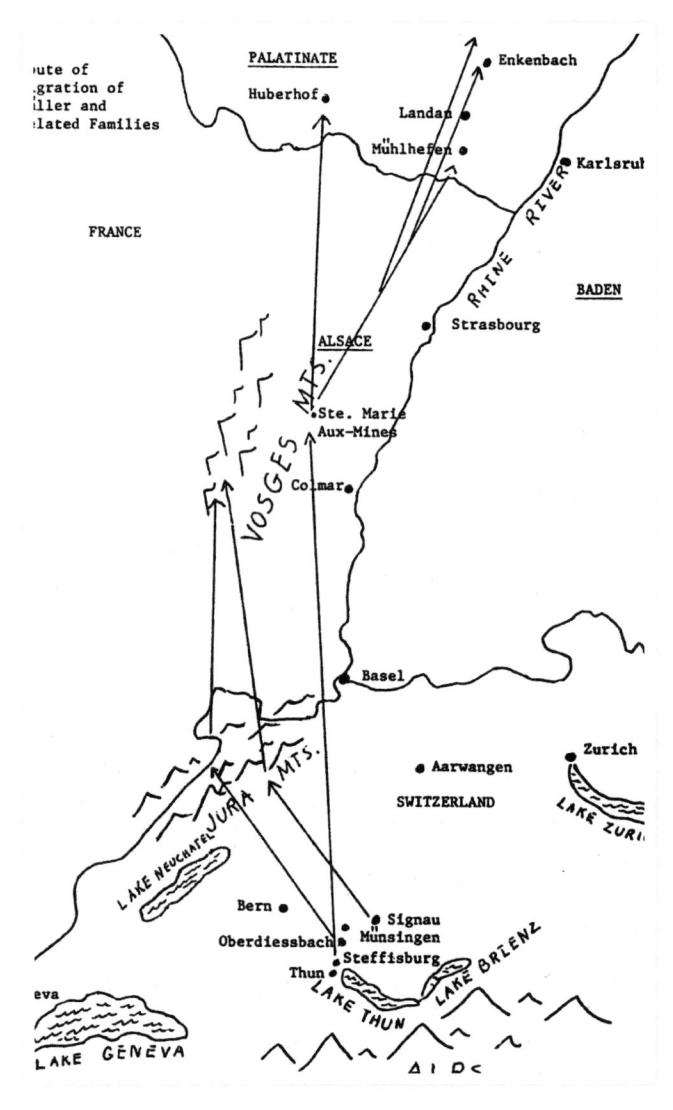

Heinrich Mueller was from Munsingen, baptized there in 1669, He was married to Magdalena Steinmann in 1693. He became a Taufer (Anabaptist) and along with many others, went to Alsace to join a group under the leadership of Jakob Ammann in 1697. He appears on a list of Anabaptists in a document preserved in the archives of Colmar, Alsace.

Heinrich Mueller's sister, Barbara married Hans Gnaegi, and they were the parents of Christian Gnaegi born in 1698. There was a brother Hans Mueller, born in 1667, and a sister Magdalena, born in 1670, these were offspring of Christen and Elsbeth (Frey) Mueller.

We do not know anything more about Hans Mueller, the brother of Heinrich. But we do know that a Hans Mueller was living at Ste. Marie –aux-Mines, or Markirch in Germany, with the group under the leadership of Jakob Ammann in 1703. Hans may have stayed only for a short time.

The Markirch group became the nucleus for the Amish-Mennonite group, that had a much stricter interpretation of Anabaptist beliefs than the other Taufers in Switzerland. The Mueller's who were our ancestors belonged to that persuasion.

The Amish next appear in Southern Germany, where Hans Mueller and Hans Wittwer came from Markirch in1699. Whether this is the same Hans Mueller who was at Marirch in 1703 cannot be proved, but there was only one person there by the name of Mueller. In 1717 Nikel and Jakob Mueller were listed as Mennonites living in the district of Alzey in the Palatinate. In the census of Mennonites that is preserved in the state archives at Karlsruhe. Claus Mueller appears also as a renter with Jakob Kurtz at Muhlofen south of Landau in 1714.

Another Jakob Mueller, who lived at Munsterhof near Dresden, also in the Alzey region. Was born in 1749 at Morzheim near Muhlhofen. He was the son of Jakob and Elizabeth (Schenek) Mueller. His father may have been Jakob Mueller who lived at Alzey in the Palatine in 1717. At Walsheim north of Landau was Christian Mueller, who was there before 1743, census lists show that he had moved away.

At this point things become difficult, it seems there are three versions about Samuel (Mueller) Miller and his children. The first is found in *Descendants of Jacob Hochstetler, (Harvey Hochstetler) pg. 959* Says that John "Broad Run" Miller was one of three brothers John, Christian and Isaac (maybe more) whose father or grandfather was an Irish boy, who at the death of his mother was placed in the care of an Amish family who raised him. This kindness to a motherless boy has brought into the Amish church many substantial families. (My father Jake used to relate this story many times)

The second story (and most unlikely story, from) *John J. and Mary Miller*

Family, (Gertie Miller), says that Samuel Mueller b.1736 came from the Canton Berne, Switzerland , and emigrated to North America on the ship "Chance" from Rotterdam to Philadelphia on Nov. 1,1763 (Samuel would have been 27 years old, and son Christian 1763-1847 was born on the same year?) When they arrived in the new world they were extremely poor, their first abode was near Germantown, Pennsylvania.

This Samuel Mueller family later moved to Kishacoquillas Valley in Mifflin County, Pennsylvania, where he became employed on the Pennsylvania canal as a teamster on the tow path, Later during the Revolutionary War Samuel was taken from his work by a group of soldiers. It isn't known whether he served in the Continental the Army or was held prisoner by the British.

During his absence his wife became very ill. While near death, she became concerned about well being of her five children. She observed the quiet, peaceful life of the Amish Mennonites and desired to have her children raised by those people. When Samuel later returned to find that his wife had died, he remarried to a non-Mennonite woman.

The third and most likely story, by researching other family books and comparing birth dates and ship lists, such as *Amish and Mennonite Genealogies, (Gingerich and Krider), Miller Family History (Eli S. Miller and Marie Kauffman, The Story of A Miller, descendants of Benedict Miller and Rachel Mast (Ken J. Heeter), Phoenix ship list 15 Sept. 1749. Thus;*

The Ship Phoenix

Christian Miller 1708-1785 (@77) m. __ Gnagi,
        On ship Phoenix 1749 (@41).

Children:

Veronica Miller- m. (Benedict) Lehman 1729-1810 (d.81)

John (Hannes) "Indian John" Miller Sr. 1730-1798 (d.68) m. Magdalena Lehman Somerset County, Pennsylvania.   On ship Phoenix 1749 (@19)

Christian Miller 1734-1777 (d.43) m. __ Mishler.

On ship Phoenix 1749 (@15)

Samuel C. (Mueller ?) Miller 1736- 1785 (d.49) m. Barbara __

Bern Twp. Berks County, Pennsylvania.   On ship Phoenix 1749 (@13)

Isaac Miller 1738-1785 (d.47)

Peter Miller 1740-__ m. Anne Zug Cumru Twp. Berks County Pennsylvania.

On ship Phoenix 1749 (@9)

Abraham Miller 1742-1812 (d.70) m. Nancy __ Conemaugh Twp. Somerset County Pennsylvania.

Nicholas __ – (d.)1784 m. Barbara STEHLY

(the numbers behind their birth and death is their age at death and the numbers behind ship Phoenix is their age in 1749).

With comparing various family records, emigrant ship lists and cemeteries (findagrave) I came to this conclusion, that most Miller families with Pennsylvania back ground can trace their ancestry back to either Christian (Schmidt), "Broad Run" John or their Uncle John (Hannes)

"Indian John Miller or "Wounded John", (He was fired upon and wounded by the Indians as they left the scene of the Hochstetler massacre.)

Comparing the father and six children's birth dates it seems reasonable this would be one family. Most children and women were not required to "sign" ship lists, and no other information is given, just their names and sometimes their age. That may have been another Peter Miller as he would have been only 9 years old.

Christian Miller 1708-1785 m. ____ Gnagi lived first in Bern Twp. Berks County, Pennsylvania. There was a land transaction in1771 in which a tract of 141 acres in Cumru Twp. Was granted to Samuel and Peter Miller. In 1772 Samuel and Barbara Miller, of Bern Twp., deeded the tract to his brother Peter Miller of Cumru Twp. In 1777 Samuel (@28) was deeded another tract in Bern

Twp., by Peter Mauer. Samuel continues on the Bern Twp. Tax lists until 1785.

Our interest with this family is with Samuel Miller's family.

Quite a bit of folklore has sprung up in connection with his three sons, Christian, John and Isaac and two daughters, Rachel and Barbara. They are said to have been orphans, children or grandchildren of an Irishman who was raised by an Amish family. This kind of story has circulated among other families. As matter of fact, the children may have been orphans in a real sense. The oldest son Christian, often called "Schmidt" Miller, was born about 1764, thus he would have been barely 21 if his father died in 1785. John, sometimes called "Broad Run" John Miller was a bit younger, born about 1766. the other children were correspondingly younger. Thus they really were orphans, and the mother was left to care for the five children, this would make Samuel born about 1736. The family of Samuel and Barbara Miller is thought to include;

Amish emigrants

Christian (Schmidt) Miller 1763-1845, m. Magdalena Berkey, daughter of Jacob Berkey.

John (Broad Run) Miller 1765-1840, m. Catherine Yoder.

Barbara Miller, m. Jacob Berkey.

Rachel Miller, m. Andrew Ochsenrider.

Isaac Miller, m. Magdalena (Handrich) Christner.

The last three left no known descendants.

BOOK FIVE

Christian (Schmidt) Miller

*Christian, the oldest son of Samuel (Mueller) Miller was received into the home of Christopher Beiler, a pioneer Bishop of Mifflin Co. Pa. Christopher being a blacksmith by trade, Christian an industrious lad aspired to the trade, in which he became quite proficient as a welder of iron, and won for himself considerable popularity and became known as *Smith Miller.

After he attained the age of twenty-one, he started to peddle household notions for a livelihood, and traveled westward across the Allegheny mountains until he reached Somerset Co. Pa. Where he purchased a tract of land from the Shawnee Indians, consisting of between four and five hundred acres. Here he married a young girl by the name of Magdalena Berkey, and proceeded to erect a cabin in the forests of Conemaugh Township. There were four white families in the township at that time, as we can imagine, their first dwelling was not an elaborate structure. But a crude log cabin closed by hanging a blanket over the doorway.

Christian Schmidt Wagon Peddler wagon

During the first summer in this typical pioneer dwelling he laid the foundation for a new home, a place where his ten children were born, and grew to young manhood and womanhood. (About five miles south of Johnstown, Conemaugh township, Somerset County) Here Christian Miller, had erected a blacksmith shop and farmers came for miles around to have "Smith Miller", repair and build implements, several days were required for some in covering the distance to and from the shop.

The wife at home, had to take a path through the woods for miles, to bring the cows home from pasture, and she heard the howling of the wolf, also met Indians in search of game.

Christian, was the first Pioneer Minister and Bishop in the Amish Mennonite church in Conemaugh Township, in Somerset Co. Pa.

He had 6 sons and 4 Daughters. Christian married as his second wife, Fannie Miller. in the year 1845. she died in 1847 at the age of 84 years. Christian and his wives are buried in the family burial ground, on a hill, close by. (South of Jamestown Pa..)

Information from "Daniel J. Hochstetler and Barbara C. Miller by Dan A. Hochstetler" pg 68.

Christian Schmidt Blacksmith, 1700s Pennsylvania Blacksmith

Dan A. Hochstetler with the Historical Society of Conemaugh Township, Davidsville, Somerset County. Penn. In which the "Schmidt" Miller farm and cemetery is located. In June of 1987 they decided to restore the cemetery which is located on the original farm, several hundred feet behind the barn up a hill on a level spot. Years ago already it was used as a pasture field and by the year of 1925 all but one of the tombstones were broken off by the livestock and laid close by on a pile. Some time later in the same year, a Miller relative of Holmes County took those stones along home to save them. Sometime during the summer of 1987 these same stones were brought back from Ohio again.

One day in June of 1987 they decided to restore the cemetery, with the help of a car load of Amish from Somerset County (the Meyersdale area) the local relatives and members of the Historical Society, the removed a temporary turkey shelter, trees and brush, hauled in top soil to fill and then leveled off the surface and then later built more or less a maintenance free wooden fence around the cemetery.

On Oct. 2nd, 1989 I (Dan A. Hochstetler) called Merle Yoder of the Jerome, Pa, a member of the Historical Society, He stated that he had the Shetler Monument Co. come and erect the original tomb stones, which were regular cut field stones with the initials and dates.

On Oct. 3, 1989 I (Dan A. Hochstetler) went to visit Merle Yoder and his wife They showed us a plate and a "show towel from the Christian Miller family" that was made to hang on a door in the inside of a home, which was custom years ago, This towel was made of linen grown on their farm. The towel had one name on it, Magdalena b. July 29, 1803 made when she was 13 years old. It reads as follows; MATDI Miler 1816.

Then We went to the (Schmidt) Miller farm, still owned by one of the descendants of the Miller family, (Mrs. (Henry) Mary Yoder). There were none of the original buildings there anymore. We walked back of the barn up a hill several hundred feet and on a rather level spot was the grave yard located. We stood on the back side of the fence, facing toward the barn.

Christian's tombstone was on the left, with his initials C M 1845 The next stone to the right was his first wife Magdalena (Berkey) Miller, With L B , and the date unreadable. The next stone to the right was his Second wife, Fannie (Veronica) Miller, no initials, Round and oblong, stones were still there in the original location, so this helped them a great deal to find the locations of the footing where the head stones had been set, With the tombstone of his first wife still standing, they placed Christian's stone on the left, And when they dug down to find the foundation of the setting of his second wife's stone, they found it had been broken off just above the footing, and when they checked the bottom surface of the stone they had left, it matched the irregular surface they had found.

Christian Schmidt Miller stone

They think their two sons, Solomon and Christian, which both died young, may have also been buried there with no markers.

*Christian (Schmidt) Miller b.1763, d.1845 m. Magdalena Berkley m. 2nd Fannie Miller <u>Somerset Co. Pa.</u>

Children;

Henry, b. May 10, 1785, d. July 31,1850 m. Anna Lehman, b 1789. They lived near Charm Ohio

Elizabeth b. 5/26/1787 d. 1/6/1816 m. Christian Joder

*Jonathan, b. Mar. 9, 1789, d. Sept. 18, 1867 m. Magdalena Kauffman They lived between Charm and Doughty Creek m. 2nd to Elizabeth Farmwald b. 11/10/1789

Solomon b. 4/23/1790 died young

Feronica b. ? /1791 d. ?/1860 m. Christian Wengard (Who came to America in 1801)

Daniel b. 8/26/1793 d. 9/8/1861 m. Rebecca Kauffman Charm/ Doughty Creek Oh. (where they had an oil mill)* (Flax oil Mill)

Jacob S. b. 10/10/1795 d. 1/26/1874 m. Catherine Kiem

Barbara b. 6/28/1797 d. ?/? m. Jacob Kauffman

Christian b. 11/22/1799 died young

Magdalena (Matdi) b. 7/29/1803 d. ?/? m. Michael Troyer (Bro. of Maria, wife of Levi "Leff" Miller)

NEW DISCOVERIES, BY D.A. HOCHSTETLER

With some further checking with some of the members of the Miller descendents, I have located a small hammer made by Christian (Schmidt) Miller and is now in the possession of a sixth generation family living in Elkhart County, Ind. We do know with his reputation he surely would have made his own hammers and may have some for all his sons. However, this hammer would have been handed down from his second son, Jonathan, our ancestor. I also found a great-grandson of "Schimdt" Miller, who was also a blacksmith and some times was called a mechanic as he operated his own Wood Work Shop and Blacksmith Shop with Wind Power, also his Planer and Rip Saws. So when they had windy weather the farmers would come for miles with lumber to have planed, as they all knew the planer would require more wind to power it, He also made a wood frame Shovel Plow with a three point steel shovel for each of his six sons, and I presume as they were married, and the youngest saved his plow and handed it down to his seventh son, and he still has it in his possession.

NEW FINDINGS OF CHRIS J. MILLER, 2016

I recently discovered who has the small hammer made by "Schmidt" Miller. It is still in the families possession. still in the sixth generation Miller family!! and I was able to take photos of it to add to my book. It thought it looked almost too good for a hand made hammer, but looking closer and doing some measurements, I can see that the hole is not in the center and other measurements are not correct as they would be on a "boughten" hammer, looking closer you can see hammer marks made when it was forged. Amazing how well this was made in a blacksmith shop around 1800 and is now prized as a family heirloom.

(My Grandfather Joni Miller showed me a knife that Christian "Schimdt" gave him one time when he was here at Holmes County to visit) (Emanuel Miller)

I wonder how many other items made by our ancestor are in existence today.

It is also said that Christian "Schimdt" Miller rode a horse, several times from Somerset County, Pennsylvania to Holmes County, Ohio and back to visit, it would have been approximately 185 Miles one way, some journey!!! He would have had at least three sons, a brother, "Broad Run" John, several nieces and nephews, and maybe other relatives in Holmes County, It was probably well worth the trip!!

In his offspring of five generations, (in 1924) He had the following Ministerial

Statistics of Twenty-One Bishops, Fifty-Nine Ministers, and Twenty Deacons. They all upheld the same banner of faith that Christian (Smith) did!

Lord, pardon what we have been,

sanctify what we are,

ORDER WHAT WE SHALL BE,

that Thine may be the glory,

and ours the eternal salvation. Amen

BOOK SIX

Jonathan Miller

Jonathan Miller, the third child of Christian (Schmidt) and Magdalena (Berkey) Miller, b .Mar. 9,1789 in Somerset County, Pennsylvania, He married Magdalena Kauffman and moved to Holmes County, Ohio in 1818, Between Charm and Doughty Creek (see photo of farm). Around 1820 Jonathan was operating a grist mill on the west bank of Doughty Creek north of TR 123. when a new road, (now St. Rd. 557), reviewed by the commissioners of Holmes County, it was to extend to " Jonathan Miller's mill on Doughty's Fork of the Killbuck". He also had an early sawmill in this location. The recognized sawmill builder Jonathan Miller, his father was the widely known Christian (Schmidt) Miller, a blacksmith by trade, was undoubtedly instrumental in the son's mechanical ability. Jonathan's brother, Daniel (Olich) (oily) also had a mill in the Charm vicinity – a flax seed oil extracting mill! He must have been quite involved with this operation to have the name Olich Daniel inscribed on his tombstone.

Jonathan's Doughty Creek sawmill was still operating at the turn of the century (1900). This mill had a vertical jigsaw- type blade, and sawing was a slow process. It was said that they had a cot at the mill to take a nap while making a long cut on a log. This water wheel setup was one of the numerous dams washed out during the flood of 1911. At the time, the mill was also washed downstream, and part of the building came to rest on top of a nearby bridge. Today the level spot and some indication of a dam wall along the creek bank are the only reminders of the once prominent place of business.

Jonathan also operated a threshing rig along with a D.D. Miller and John Frey. From about 1837 to 1875. A threshing rig usually consisted of a steam engine (at that time), a threshing machine, a water wagon and coal or fuel wagon, and traveled from farm to farm to do threshing through out the neighborhood.

In the year 1866, widower Jonathan Miller moved to Lagrange County, Indiana, with his son, Joni and family. It is said that his Nephew Christian C. Miller (my Grandfather) went with a horse and buggy to Charm, Ohio and took Jonathan along back to Lagrange, County, In. Jonathan was 77 years old. He purchased a forty-six acre farm, on County Road 300 west at the corner of 400 South. They first lived in a small plank house, which was later moved to a neighboring home, and was at one time used for a wood shed. In Jonathan's older years he told his son, Joni, "When I die." Pointing down the road, "Bury Me under that old Cherry tree," which they did. He died Sept.18,1867, and was the first one to be buried in the now well known Miller Amish Cemetery. County Road 300W South of 350S

Typical Threshing Rig

Mystery solved, by Chris, In Ohio it was thought that Jonathan was buried at another cemetery near one of his other children. I was surprised when I discovered he was buried in Lagrange County, Indiana, In Indiana it was thought that Magdalena was buried in an unmarked grave.

Jonathan Miller's Grave at Miller Amish Cemetery, Lagrange Co, In.

Magdalena's Grave at Cemetery O-1 John B. Yoder Farm N. E. of Charm, Ohio

Jonathan and Magdalena (Kauffman) Miller, Their Children:

Feronica b. Oct.13, 1806 d. Feb.3,1874 m. Jonas Miller m. 2nd to Paul Hershberger

Marie b. Feb. 23, 1809 d. Aug. 2, 1869 m. Joseph Christner. Minister, Amish, Lagrange, In.

Barbara b. Oct. 3, 1810 d. Jan. 21, 1889 m. Michael Troyer (bro. of Maria, wife of Levi "Leff" Miller) m. 2nd Benjamin Gerber

Elizabeth b. Jan 29, 1812 m. Jacob Gerber

Magdalena b. Aug. 3, 1813 d. Aug 20, 1875 m. Elias Hershberger, Sugarcreek, Oh. Amish

Gertrude, b. July 30, 1815 d. Mar. 13, 1858

Jeremias b. Aug. 15, 1817 d. Jan. 1890 m. Lydia Troyer, Martins Creek and Charm, Ohio.

Anna b. April 22,1820 d. Oct. 21, 1899 m. Simon Miller m. 2nd to Bishop Shem Miller, Amish, they lived near Doughty Creek.

*Christian J. (Christal) b. Aug 31,1822 d. June 23, 1900 m. Catherine Frye m. 2nd. Barbara Stutzman, Farmerstown, Ohio, Amish.

Joni b. Sept. 1,1824 d. Aug. 24,1913 m. Susan Hochstetler.

Benedict b. Jan. 11,1828 d. Aug 22, 1916 m. Rachel Mast, Butte Creek, Clackamas County, Oregon.

Jacob b. Sept. 22, 1830 d. Feb. 25, 1908, m. Mariah Mast m. 2nd Sarah Greiner. Indiana, Mennonite.

Jonathan Miller farm, Charm, Ohio

This farm, located about two miles north of Charm, Ohio, was owned by our ancestor Jonathan Miller and his son Christian for about forty years. Note the cemetery in upper right where Jonathan's wife Magdalena is buried but Jonathan is buried in Lagrange County, Indiana in the Miller Amish cemetery.

Right near the highway, SR 557, and guardrail lies Doughty Creek (hidden), several feet beyond is where the flax mill was located, operated by Jonathan and his brother "Olich (oily)" Daniel and powered by a water wheel. It would have produced linseed oil which at that time would have been used as a finish and preservative for furniture and other wood products.

Stones of foundation where flax mill stood.

The Doughty, "Troyer Hollow", today is a private park now used for weddings, reunions and picnics. At one time it was a bustling industrial area

44   A JOURNEY TO THE FUTURE

Doughty Creek used for water power

SR 557 on right Charm Engine in background

The Doughty

with several mills and a dam to supply water power At one time there were about three mills in the general location of the building above. There were at least three homes in that area where people lived until about 1942, when the last residents moved out.

The narrow valley is prone to flooding and occasionally got wiped out by flash floods and is not now suitable for habitation, but it is a very pleasant valley with Doughty Creek running the full length and some caves along the rock walls.

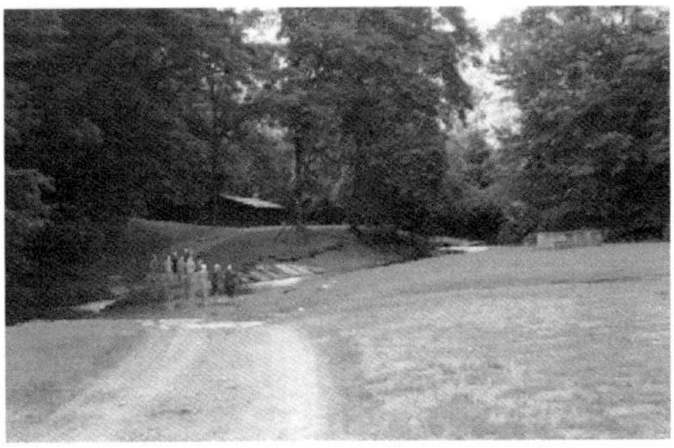

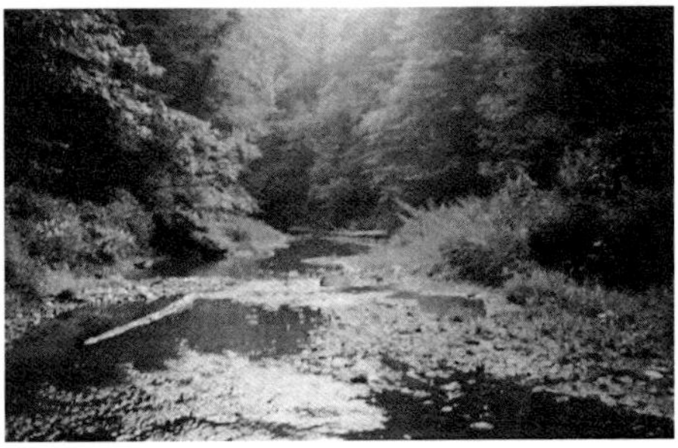

BOOK SEVEN

Christian (Christal) J. Miller

* Christian (Christal) J. Miller b. Aug. 31, 1822 d. Jun. 22,1900 m. Jan. 22,1843 to Catherine (Katie) J. Fry b. Feb. 20,1828 d. April 8, 1877 both born in Holmes County, Oh.

They owned and occupied a farm one half mile south and one fourth mile east of Farmerstown, Oh., Before that they owned a farm three fourth mile north and ½ mile west of Charm,Oh., That farm was bought by his father, Jonathan, in 1831, who owned it for thirty years. In 1875 it was owned by Christian J., which he purchased it for his son, Daniel C. Miller. The Cemetery on this farm contains fourteen graves, including Magdalena, wife of Jonathan. All the rest of those buried there are descendants or in-laws of Christian and Catherine.

This is the farm south of Farmerstown, Oh. where "Christal" lived and raised his family. In the original arrangement the house stood on one side of the road and the barn on the other, but with increased traffic it was not practical. Later on someone built a large house on the same side as the barn. (lower right)

The original barn was hit by lightning and burned about 1935-6, and was rebuilt on the same foundation and a straw shed added later. The old house (two story in center of picture) was razed after a new house was built to the left. That made two homesteads across the road from each other.

This farm is located about a mile southeast of Farmerstown, Oh. on St. Rd. 557 and today is occupied by the Allen J. Miller family.

Early maps show that Route 557 went diagonally through the center of the original farm.

Catherine died April 8, 1878 and is buried near Farmerstown, Oh. Aged 51 yrs 1 mo. 18 days. Christian married second time to Barbara Stutzman, June 4th 1878, She was born April 11, 1824, in Holmes County Oh. and died Oct. 10, 1896, at the age of 72 yrs. Christian and his two wives are buried side by side with Christian in the middle, near Farmerstown, Oh.

Catherine  *Christian*  *Barbara*

The stones for Christian and His two wives Catherine and Barbara are written in German and located in cemetery O-21 on the Andrew Yoder farm, ½ Mile south and ¼ mile east of Farmerstown, Oh. Behind the Wicker Furniture store.

So far I have not been able to get any more information about Christian J. Usually I can find stories or happenings about people, but so far nothing, but

I'll keep trying.

Christian (Christal) and Catherine, their children;

Elizabeth 1844- 1846

John C. b. Sept. 25, 1845 m. Fannie Hostetler Haven, Ks. (Alice's ancestor)

Barbara C. b. May 27, 1847 m. Daniel Hostetler Emma, In. (father of Daniel J. Hochstetler "Amish Historian")

Benedict C. b. Sept. 24,1848 m Lizzie Hershberger m. 2nd to Elizabeth Troyer Charm, Oh.

Jonathan C. b. Dec. 30, 1849 m. Anna n. Miller (She died) his 2nd wife was Elizabeth Shetler widow of Isaac J. Yoder, Sugarcreek, Oh.

Daniel C. b. Oct. 3, 1851 m. Catherine Hershberger (She died) his 2nd wife was Susan P. Schrock (Daniel's first wife and Benedicts were sisters.) Charm, Oh.

Martha C. b. Mar. 27, 1854 m. Abraham Raber Baltic,Oh.

*Christian C. b. Aug. 2, 1856 m. Mary Bender Lagrange, In. (my Grandfather)

Catherine C. b. Nov. 27, 1858 m. Moses B. Beachy Walnut Creek, Oh.

Mannelius (Neal) C. b. Feb. 20,1861 m. Susan E. Miller Farmerstown, Oh.

Mary C. b. Apr. 24, 1863 m. Mose Miller Walnut Creek, Oh.

Samuel C. b. Apr 27, 1866 m. Sarah Raber Benton, Oh.

And the peace of God
WHICH PASSETH ALL UNDERSTANDING SHALL
keep your hearts and minds through Christ Jesus.

Phil. 4-7

BOOK EIGHT

Christian C. Miller
"Iowa Christ"

Christian C. Miller b. Aug. 2, 1856 in Holmes Co. Oh. D. Mar. 22, 1923, m. Nov. 5, 1878 to Mary Bender b. Dec. 21, 1856 in Somerset, Pa. Dau. Of Michael and Catherine Bender who lived near Kalona, Ia. at the time of Christian and Mary's marriage, Nov. 1878. Due to this that they had lived in Iowa for awhile, He was known as "Iowa Christ" He was ordained a Deacon in the Amish Church. He was also a cabinet maker, thresher and farmer.

They lived at "Clearspring" Lagrange County, In. 3640W 350S where a grand-daughter now lives in the Dawdy house. East of the farm on CR. 300W is the Miller Amish Cemetery, where Jonathan Miller (the first person buried there) and my grandparents, Christian C. and Mary (Bender) are buried plus two young sons, Sylvanus and Ammon and three daughters, Mattie, (Mrs. Dan

B. Miller), Elmina, (Mrs. Oba J. Miller), and Lydia, (Mrs. Jacob Schlabach).

My Mom (Ada) grew up on this farm and went to a nearby (Taylor) one room school, Mom always called this their Homestead, She lost two of her brothers when they were only teenagers, it must have been sad time for her as she talked about them a lot. When I was growing up, Jacob Schlabach's lived on the homestead, Mom's sister Lydia was married to Jacob at one time. Although Jacob was married the third time and had a total of twenty children and step children, I used to think they were all my cousins, but come find out only six were my actual cousins, Olen, Neomah, and Ida Mae Miller and Alvin, Walter and Mary Ellen Schlabach.

We used to go visiting Jacob Schlabach's a lot. (about 1936), And I remember coming home late at night with my parents and siblings in a double-buggy, with two horses clippity-clopping during the night and watching the airplane beacon light about half way home, watching the stars and moon, the only light we had on the buggy was a kerosene lantern with a red lens toward the back, but we hardly ever met a car, we lived north of Millersburg at the time, about 10-12 miles to our house.

At that time there were two big houses, where the smaller house is now, there was one almost as big and there was a porch clear across the front of both houses.

Christian's wife Mary died in 1899 and Christian married 2nd. To Barbara (Hershberger) Yoder (widow of Jacob M. Yoder, Berlin, Oh. They had a still born son b.1901 Barbara brought along five children,

(My Mom often talked fondly about her step-siblings) they would have been close to their same age.

I noticed that Barbara was not buried beside Christian C. so I did some research and found that she is buried beside her first husband, Jacob M. Yoder in cemetery K-5 on the Moses Coblentz farm west of Berlin, Oh. north of U.S. 62.

In Olen's Story he says that when my sister Mattie and Elmer K. Miller were married, on Dec. 22, 1932, that afternoon, Mom, Dad and Mom's sister and husband, Ben-Dan and Mattie Miller went to Barbara's funeral in Holmes county, Ohio.

Amish Miller Cemetery Cr. 300W south of 350S.

Christian C. Miller
Both buried in the
Miller Amish Cemetery,
Lagrange County Indiana

Mary (Bender) Miller

The Christian C. Miller Homestead.

Children;

Mattie b. Oct.9, 1879 d. Sept. 2, 1943 m. Daniel B. (Ben-Dan) Miller Goshen, In.

Sylvanus b. Feb.7,1882 d. Sept. 11,1887

Ada and Katie b. July19, 1886 *Ada d. Feb. 18, 1963 m. Jan. 2, *Jacob J.C. Miller Millersburg, In.

Katie b. July 19, 1886 m. Enos Bontrager Shipshewana, In. d. Feb. 15, 1919 she is buried with an infant in her arms.

Elmina b, July 10, 1889 d. Nov. 8, 1918 m. Oba J. Miller Amboy, In. (Dad's brother)

Lydia Ann b, May 30, 1892 d, July 22, 1933 m. Ervin R. Miller – Lydia m. 2nd to Jacob Schlabach, Lagrange, In.

Ammon b. May 25, 1897 d. Aug. 31, 1910

Mary (Bender) d. Mar, 10 1899. Christian m. 2nd to Barbara Yoder they had one child not living.

Barbara was m. 1st to Jacob M. Yoder Their children;

Abraham b, 1885 1st. M. Barbara L.Yoder 2nd to Anna Hershberger 3rd. Barbara (Miller) Kauffman, Topeka, In.

Mary b. 1887 m. Albert Wengard, Fredericksburg, Oh.

Levi b. 1888 m. Emma Yoder, Beach City, Oh.

John b. 1891 m. Mary Miller Wyoming, Del.

Susie b. 1893 m. Jacob Lambright, Lagrange, In.

(Susie died May 3, 1911, Jacob then married Dad's sister, Lizzie b. 1893 d. 1918, Daughter of John J.L. and Mary Miller.)

HISTORY OF THE BENDER ANCESTORS OF (MARY BENDER)

German-Swiss house

Among the names in the Steffiburg of the Berness Oberland Area, in the late 1600s to 1700s, we find the Bender's along with many present Amish surnames such as, Miller, Eicher, Graber, Christener, Hofstetler, Joder, Mast, and many others. The name Bender also shows up in the Alsace-Palatinate area working on farms/estates of Prince Friederich's Billinghauserhof dairy farm around 1732.

Daniel Bender, Langendorf, Germany, was born in Germany, He Married Magdalena Schlappach who died in Germany. Daniel died in about 1842 and was buried in Germany. They had 3 children namely, Christena, Daniel and Barbara. Daniel was married the second time in Germany to Elizabeth Bauman.

The children of Daniel and Elizabeth (Bauman) Bender were; Wilhelm,

Joseph, Christian, Johannes, Jacob and *Michael.

It was decided Wilhelm, the oldest of the second marriage, should go to America before he reached the age when young men must enter military training. But the family was very poor, passage to America was beyond their means! Wilhelm became a redemptioner. Peter Kinsinger, an Amish friend who was coming to America, paid Wilhelm's passage. When they landed at Baltimore, they came in contact with the proprietor of a nursery, who paid to Kinsinger the amount of Wilhelm's passage, with the understanding that he would stay as a redemptioner, and work out the sum he had paid for him.

Kinsinger came west on the National Trail and joined the Amish settlement at Somerset Co. Pa. while Wilhelm, a boy of about 15 years of age was left near Baltimore.

When a plan to rescue the stranded Amish boy was discussed in Bishop Benedict Miller's home in Somerset County, Miller's 19 year old daughter Catherine suggested that "he might someday make a good husband for someone" Bishop Miller rode to Philadelphia, paid the boy's redemption fee, and brought him back on horseback. Eight or nine years later Wilhelm and Catherine were married.

By hard work and saving, he earned enough money to pay for his brother Joseph's passage to America. Later in Germany sorrow had come to the Bender home. Shortly before they wanted to leave, the father, Daniel died, but the sons in America did not know this. The widowed second wife of Daniel Bender, Elizabeth Bauman, aged 48 arrived at Baltimore on the

Bark Elizabeth Hall

Bark "Elizabeth Hall" on August, 23rd. 1842, accompanied by her four youngest sons, Aged 11, 15, 18, and 20. Her two oldest sons, Wilhelm and Joseph had arrived separately, in the years 1829 and 1837. Wilhelm was now married and had established a home. Because the Daniel Bender family was very poor and lacked means to raise money for the ocean passage, the two brothers in America faithfully saved their earnings so that by 1842 they could provide for the immigration of their parents and four younger brothers.

After their arrival in Baltimore the widow's family had only enough money left to pay for stage coach to Cumberland. The last twenty-seven miles to Wilhelm's home near Grantsville was then accomplished on foot. The foot weary travelers were provided with potatoes for supper and bread for breakfast by a kind man in whose barn they spent the night along the way.

A family tradition tells of how the older son Wilhelm, in anticipation of the family's arrival, had fattened a calf for the occasion . The day came when the calf was slaughtered for the family's use. On that same day the Bender's saw a group of people approaching from the direction of the National Trail. Wilhelm remarked "They look like my family, but there is one too few" When they arrived they were his dear Mother, and four younger brothers but his father was not along. Thus for the first time Wilhelm and Joseph learned that their Father was dead. This was in 1842

* Michael Bender the youngest son, was born in Germany Nov. 11, 1830. He married Oct. 17, 1852, by Michael Hay Esp. to Catherine Hostetler, born April 10, 1828 near Summit Mills, Pa. Catherine died May 10, 1874, near Kalona, Ia. And buried in Lower Deer Creek Cemetery. With her he had the following children:

John M. m. Savilla Keim, Elder, Judge, Presbyterian, Imperial NE.
*Mary M. m. Christian C. Miller, Cabinet maker/farmer Amish Topeka, In.
Samuel M. m. Saloma Yoder, Farmer, Menn. Wellman,Ia.
Emma 1861-1868
Joseph 1865-1882
Daniel M. m. Lethie Wood, Methodist. Burlington Ks.
Katie m. Charles Murphey, farmer, Protestent. Iola Ks.
Lydia m. Peter B. Nichol, Hartford Ks.
*Michael married 2nd to Susanne (Kaufman) Hostetler, widow of Noah J.

Hostetler of Grovertown, In.

Their children;

Eli m. Maude Elliott, Harper, Ks.

Fannie Pricilla m. Wood Gillet, Aurora Mo. m. 2nd. to Charles R. Wolf Marionville, Mo. m. 3rd. Andrew Homer Calhoun, Merrill, Ia.

William m. Idella Corey, Wayland, Ia. Carpenter, Methodist.

Margaret Edith m. Haro J. Schake, Durango, Co. Director of Markets, Presbyterian.

Susanne had with her first husband, Noah J. Hostetler, the following children:

Sarah m. Isaac Slabaugh, Moses m. Rebecca Waybrant, Levi m. Cora (Shetler), Mary b. 1867 d. 1868, Susanna b. 1869 d. 1869, Noah b. 1870 d. 1872, Elizabeth m. Emanuel Stutzman, Emma m. Lemuel Kankford.

Not being able to find out any stories about Michael Bender, I find his family genealogy rather interesting, to see the blending of three families and who the children married and where all the places that they settled.

Michael must have had an interesting life, just by tracing his foot steps: Born in Germany, he crossed the Atlantic in 1842 on board the Bark "Elizabeth Hall" arriving in Baltimore, Md. Traveling with his widowed mother and three brothers, then making their way overland on the National Trail by stage coach and walking to his brother Wilhelm's place near Grantsville Maryland. Growing to adulthood in Pennsylvania and finding and marrying his first wife, Catherine, near Summit Mills, Pa. Then to Kalona, Iowa where he buried Catherine in Lower Deer Creek Cemetery, Leaving him with six children. And his daughter Mary finds a husband, Christian C. Miller. Michael finds his second wife, Susanna, a widow with six children from Grovertown, In., and then they had four more children!

I found tombstones of Micheal and Susanna on findagrave.com they are both buried in Hartford Cemetery, Near Hartford, Lyon County, Kansas. Along with 17 yr. old son Joseph. Micheal died almost half a world away from where he was born.

Michael *Susanna*

What did he look like, tall, short and what was his disposition? Timid, Kind, Surely Adventurous, a pioneer looking for new lands to conquer! How did he travel?

Various ways that he could have traveled are by sailing ship, two wheel cart, walking, Conestoga wagon, prairie schooner, stage coach, canal boat, and early steam train. But chances are he walked much of the way, as riding was reserved for the very young or too old and the frail.

A JOURNEY TO THE FUTURE

BOOK NINE

"Broad Run" John

John Miller the second son of Samuel and Barbara (Mueller)Miller, Born in Berks County, Pennsylvania about 1766. In his early twenties he married Catherine Yoder. By this time they were looking to lands beyond the Appalachians in Somerset County, where other Amish-Mennonites had begun to settle about 1779. By 1790, when the first Pennsylvania census was taken, John and Catherine were living in Elklick Township near his cousin, John Miller, who was married to Catherine's sister Freny.

John "Broad Run" Miller (DJH 9175) 1765-1840 b. Berks Co. Pa. M. 1785 Catherine Yoder b. 1765 Berks Co. Pa. Around 1810-1820 John and Catherine moved to Tuscarawaras Co. Ohio

Children; all children born in Somerset Co. Pa.

Daniel C. Miller 1789- ? m. Elizabeth Troyer b. 1800 (DJH 9176) Walnut Creek, Oh.

John Miller 1791- 1861 m. Miss Chupp, Elkhart Co., In., in 1839

Simeon/Simon Miller 1793-1874 m. Elizabeth Gnagy b. 1793

Catherine Miller 1795-1871 m. Moses "Grosse" Miller b. 1802 (DJH 9168), Holmes County, Ohio

Elizabeth Miller 1795-1865 m. Michael Miller b. 1798

*Bishop Levi "Leff" Miller 1799-1884 m. Mary Troyer b. 1806-1872, Charm, Oh.

Elias Miller "Lame Eli" Miller 1803-m. Rachel Holley, Owens County, In.

Anna Miller 1805-1892 m. Abraham Gerber b. 1804 Walnut Creek, Oh.

Barbara Miller 1807-1885 m. Henry Schrock b.1807 Trail, Oh.

These families and a few others, such as the Christners, Schrocks, and Hochstetlers formed the Elklick Amish-Mennonite church. In 1809 they wished to move further west to Ohio. Land was opening up in Tuscarawas County in the Sugarcreek Valley. In 1810 John Miller was living in Quemahoning Township in the northern part of Somerset County, Where his older brother Christian (Schmidt) and his younger brother Isaac had settled. The urge to move was still there.

By 1803 Ohio had achieved the congressional right to statehood. The first

Conestoga Wagon, drawn by Velma Peck

evidence of an interest in Ohio among the Somerset Amish was the report of a party of three who scouted the wilderness of eastern Ohio by horseback in 1807. They returned with evidently a glowing report of what they had found along the Sugar and Walnut Creeks, rolling land, a good bit more gentle than Somerset County,

Massive oak trees, an indicator of fertile soil, and bountiful springs gushing from the hillsides.

Around 1810 or 1820 John Miller and his wife with nine children aged 3 to 22 pulled stakes to move to Ohio. We don't know much how they moved, but we know others who moved at the same time. Two or three families grouped together and used a Conestoga wagon pulled by the teams of each family, the able bodied walked, while the youngest rode in the bumpy wagon, The wagon was filled with necessities such as, a large family Bible, a song book (Ausbund) plus axes, shovels, hoes, a crosscut saw, a scythe, an iron kettle, a blanket chest, a supply of salt, and a few dozen panes of window glass were considered musts. Also included were probably a spinning wheel and a plow, broad axe and adze. The axe was almost indispensable for the pioneer.

A supply of corn, wheat and other grains for seed, vegetable seeds, and a variety of fruit tree saplings. With a colt or an extra horse and a cow or two herded along provide fresh milk as they traveled, with the surplus milk hanging below the wagon and shaken as they traveled supplied them effortlessly with butter.

At that time there was no bridge or ferry at Wheeling W.Va. so they had to take a northern route, they had to ford the Allegheny river at Pittsburg and enter Ohio near East Liverpool, then on to New Philadelphia and west to Sugarcreek. Some times widening and repairing the road as they went.

Some time between 1810 and 1820 *John Miller,* took up land warrants in

Dover Township, east of Sugarcreek, Ohio, It was actually beyond the edge of the main Amish settlement, by a stream called Broad Run and as a result John there -after was called *"Broad Run" John.*

Pioneer cabin drawn by Velma Peck

I first heard this following story from Henry Erb in 1992, Then, later on I heard the same story from LeRoy Beachy, (It can be found in his two volume books "Unser Leit". *"Broad Run"* and his family settled five miles northeast of Sugarcreek. The chosen quarter-section was a hilly sloping ridge with one low bottomland corner barely touching Broad Run creek. Though the location kept his family some what isolated from the other Amish, All of his children married in the faith.

A handed down story tells of *"Broad Run" John's"* discovery, a number of years after he had settled, that the deed he held and the quarter section for which he had been paying yearly taxes on, was not for the one he was living on. The deed he had was for some low swamp land further east. Apparently the land office recorder had made the mistake entering 21 instead of 12 and since no one had as yet applied for the quarter section 21 the discrepancy had long gone unnoticed.

The morning *"Broad Run"* left for Zanesville (the land office at the time) on horseback to have the error corrected. When he stopped at the livery station in Coshocton to refresh his horse, During a conversation the station operator informed John that a neighbor who had heard of the recording error, had stopped in earlier in the day and was also headed for Zanesville. He boasted that he was on his way to lay claim to the quarter-section the Miller family had settled on and had made considerable improvements. The livery man then offered *"Broad Run"* his own horse, the fastest one in the stable, Suggesting that if he drove him hard, he could possibly pass the man on the way, *"Broad Run"* accepted the offer.

When he caught up with the neighbor on the trail *"Broad Run"* pulled his broad-rimmed hat down over his forehead, turned up his coat collar, hunched his shoulders, and sped on by, arriving at the land office well ahead of his opponent. After taking care of the error and as he walked out the door the neighbor was coming up the walkway. That's how near *"Broad Run"* almost lost his farm they had worked so hard for.

Some believed *"Broad Run"* was buried on this farm and likely they used a fieldstone marker, it may or may not have been inscribed. The farm was later strip mined and the stone and grave was likely lost at this time, which was not unusual at that time and area. The photo shows the sign "Broad Run Dairy" which is the road that passes the farm that *"Broad Run"* owned. The next photo shows the un-natural contour of the land that has been strip mined.

Later his older sons Daniel, John, and Simeon had farms between Charm and Carlisle (now Walnut Creek). In 1820 John still had two sons, Levi and Eli, and three daughters Catherine, Anna, and Barbara at home, according to the census.

"BROAD RUN" JOHN

In 1830 the children were all married except Barbara. John and Catherine were now located near or on the Moses Miller farm south of Walnut Creek (Desc. Of Eli S. Miller and Marie Kauffman 1981), Moses was married to John's daughter Catherine. Levi was married to Mary Troyer and he became an Amish bishop.

The brothers-in-law, Levi and Moses Miller, became leaders of the rival Amish-Mennonite and Old Order Amish groups in Holmes County. Moses was the leader of the progressive group, while Levi led the Old Order movement. (Levi's son John moved to Elkhart County, Indiana in 1862, known as "Leff John" Miller he also became a well known Amish bishop).

Broad Run John Miller did not live to see this division, since he died before 1840, as shown on the 1840 census of Holmes County. Apparently his widow Catherine was still living at the Moses Miller home. Her gravestone was recently found at a corner of the ancestral farm of Moses Miller. She died between 1840 and 1850. This farm is now owned by Harry Gerber Sr.

Thus the life span of Broad Run John Miller, which began in Berks County, Pennsylvania, ended near Walnut Creek, Ohio.

When we pray for rain,
we must be willing to put up with a little mud.

BOOK TEN

Bishop Levi "Leff" Miller The Furniture Maker

Levi "Leff" Miller was the son of "Broad Run" John Miller, 1799\1884 He was a Bishop in the Amish Church, Holmes County, Ohio. He lived about Two miles south-east of Charm, Ohio.

On October 5,1981, my nephew Leonard Miller, Auctioneer, bought a rocking chair at my cousin Owen Miller's sale. Owen said this chair was made by Levi "Leff" Miller in Ohio, Leonard took the chair to have it refinished by Bill and Mary Miller. When they looked at it and discovered this to be a unique built chair, it was made of walnut wood and held together with wooden pegs and is as solid as when it was made.

Leonard found out this chair was made by his great, great, great grandfather, a noted Amish Bishop.

When I first became interested in Levi "Leff" was in 1992 when Alice and I took a trip to Holmes county Ohio. Nephew Leonard had told us he thought "Leff" lived about two miles south of Charm, Ohio, We met an Atlee Schlabach at Charm, Ohio and He showed us where "Leff" had lived and also where He is buried beside his daughter, Catherine. Atlee also introduced us to Henry Erb (Amish Historian) Who gave us much information about "Leff" and also about His father "Broad Run" John Miller.

What with Leonard's chair that was made by "Leff" I started getting more interested and whenever I went to Holmes County I would inquire as much as I could. Then in Oct. 20, 2014 Leonard and I took a trip to Holmes County and spent several days to find more chairs and furniture that He had made. We were amazed at what we found, but we also began to realize that there was also a spiritual part of his life that we would like to find out more of his role as an Amish Bishop, involved with issues concerning admitting people from other churches, concerning methods of baptism and being involved in a rare Amish murder mystery.

He seemed to always have compassion, words of wisdom and feelings for other people, I will tell you more about that later, in the meantime this is what we found out about his furniture making.

Bishop Levi "Leff" had a wood working shop where he made rocking chairs and some other furniture. We learned that he used a horsepower "sweep" to drive his lathe and other machines. The horse and "sweep" were on the lower level with his wood working equipment on the floor above, it is said when he wanted a little more power he would rap on the window or floor to make the horse go faster. The upper pulleys would have been connected to a line shaft with other pulleys to drive a lathe, saws and various other equipment. The workshop was located north of the present two houses, and has long been torn down. The quest now

A JOURNEY TO THE FUTURE

is to find out how many rockers and other furniture that he made.

I have to wonder what kind of wood working equipment, what did his lathe looked like and what other equipment, saws etc, did he have. The Amish were known to have some of the latest equipment such as grain reapers and threshers. The big problem was available power, this was before steam engines were readily available and gas engines had not come into use yet.

Water power was the most common source of dependable power in Holmes county then, as there were many streams and rivers in the area and all they needed to harness the water was to build dams and build water wheels, which were quite common in the Charm/Doughty area. But for some reason "Leff" chose to use "horse power" as I described earlier, this is (see sketch) only my guess as to what it looked like, it's as close to what I could find to describe what it looked like. It is amazing that the rocking chairs and other furniture he made in the mid-eighteen hundreds are still as sturdy as when they were made.

Having been made in the 1850s, The ones we have found are extremely sturdy made and all in excellent shape except for the cane bottoms which most have been replaced, because the webbing usually does not last long, although some seem to have the original webbing. The turnings are amazing, he must have good equipment for the day, and I think he was an excellent craftsman to put out such good furniture that is now roughly one hundred and sixty years old. In the photo on right you can see one of the small wooden pegs that holds the chair together.

Inquiring what makes these rockers so unique an expert woodworker told me that (He had a secret with the pegs) (modern woodworking says pegs are not practical).

The little child's rocker has been played with by generations of children and is still solid as when it was made. The children not only rocked dolls on it but it is said some boys turned it up side-down and pushed it across the floor, as a make believe plow or what ever. If you look closely you can see the top spindle is worn on both sides to almost a taper, but it keeps on rocking.

Most of these rockers have multiple coats of paint as a lot of time it was much easier to give it a coat of paint rather than stripping and refinishing. One woman said, as her husband was removing many coats of paint, "I think the paint is the only thing holding it together" but he found out different when the paint was removed.

It seems like every one of these pieces has a story, either it was passed down from ancestors or family members. Not surprising we found none of this furniture for sale.

Levi "Leff" Miller 1790-1884 m. Mary "Maria" Troyer, dau. of John and Magdalena Troyer (One of the first Amish settlers to establish the Walnut Creek valley settlement) Mary 1806-1872 Is buried in cemetery O-18 Melvin Miller farm, (no marker and no date of death). Levi "Leff" is buried beside his daughter Catherine (Schlabach) in cemetery O-3. When "Leff" became senile in his old age, he lived with his daughter and she took care of him, that's why he is buried next to his daughter.

When first married, Levi and Mary lived on her folks (John and Magdalena Troyer's) home place, North of Walnut Creek.

The east half of the southwest quarter of section eight, on which the cemetery (O-18) is located, (3745) State Road 557, was first deeded to Levi Miller on April 1,1829.Levi remained there the rest of his active life. It

A JOURNEY TO THE FUTURE

was on this farm where "Leff" had his furniture making shop. It is the first house on left partially hidden behind trees. In 1861 the farm was owned by Levi's son Simeon, who, some time before 1875 sold it to his brother-in-law Daniel (Anna) Raber and moved to Indiana. Daniel's wife Anna died in 1875 and Daniel remarried to Lydia Stutzman (Miller), so then the people living on the farm were not blood relatives of "Leff", if this is the case then this could have been cause of some uneasiness between the them, this situation has a lot of potential for tension between the two parties.

LEVI "LEFF" MILLER FAMILY LINE

DBH 6749 Levi "Leff" Miller 1799-1884 m. Mary (Maria) Troyer b. 1806 d. 1872

Children:

Magdalena 1824-1885 m. Abraham Nisley Goshen, In.

John "Leff John" 1826-1890 m. Annie Hochstetler (DJH7699) Goshen, In.

Simeon 1829-1876 m. Lydia Miller m. 2nd to Mary Graber Goshen, In.

Catherine 1831-1906 m. David J. Schlabach Charm, Oh.

Anna 1883- 1875 m. Daniel Raber Sugarcreek, Oh. (Daniel m. 2nd to Lydia Stutzman (Miller).

Elizabeth (infant)

Sarah 1837-1863 m. Jacob D. Mast Mt. Hope, Oh.

Levi 1839-1905 m. Fanny Chupp Mt. Ayer, In.

In researching for Levi "Leff" Miller, I began to find a man with an interesting life, They must have been good parents as they raised seven children, one of them became a prominent Amish Bishop in Elkhart, County, Indiana.

"Leff" was a minister, then a Bishop in the Amish Church, his name comes up quite often in church affairs at that time, also he was a figure in an Amish murder mystery that lasted over fifty years until someone confessed to the sinful deed. A small child was suffocated by a spurned suitor. An innocent man was accused, and lived as an outcast from the church and community until almost at the end of his life the real culprit finally confessed to the horrible deed, One of the people he consulted was Bishop "Leff" Levi Miller. But by that time most of the people involved were gone. And the statute of limitations had expired so he was never prosecuted. But the innocent man lived to hear his name cleared and he expressed forgiveness for the guilty one, even thou he lived for fifty years with the accusation on his head.

A lot of controversy at the time was about baptism, it seemed lots of people had their own ideas. One was the method of baptism, baptism in running water (immersion) versus pouring\sprinkling. Here is a letter he (Bishop Levi "Leff" Miller) wrote to the church expressing his feelings.

Levi Miller, the oldest minister (Bishop) in Holmes Co. Oh. Wrote; *As we understand it the main offense concerns baptizing in the water, and Solomon Beiler does not want to baptize in the home. So we wish to remain at peace with him, since it (baptism in the water) is very close to the way Jesus Christ has gone before us.*

We certainly must bear that in love since (baptism in the water) is closer to the word and example of Jesus Christ than the old practice which was begun at the time of the martyrs, since they had to keep their divine ministry in secret. I also believe in this, for I believe the difference is minor. (Levi Miller)

"Leff" also held steadfast and held together (along with his nephew Glay Mose) the ways of the old order Amish church, despite the split where his Brother-in-law Grose Mose became leader and Bishop in the more progressive church, allowing meeting houses and automobiles etc. and still remained very good friends. The more I find about him, he is one I wish I could have met and I feel privileged to have him as one of my ancestors. He was also not afraid to etch

A JOURNEY TO THE FUTURE

his name in stone, as evidenced on this limestone rock used as foundation for a hog house/tool shed on his farm, (L.M. 1851), also in the barn, on the foundation there is a crude marking (L M) w/date (unreadable), scratched by machinery. I have to wonder just how did he do this and what tools did he use ? Considering the year it was made and the precision that it is made, as precise as a tombstone, protected from the weather it is almost as it was when engraved.

From what information I could gather I think he was an amazing man, but then when he became older what tragedy! That he became senile\dementia and was to the point where he was almost uncontrollable so that they had to lock him in his room to control him in his last years, (The Amish at that time would, if at all possible take care of the mentally ill themselves in stead of sending them to a public institution) And a loving daughter took care of him to the end and buried him in a grave beside her. In the Schlabach cemetery east of Charm, Ohio, Where at that time he was the only Miller buried there. *Uncle Chris*

LEVI "LEFF" MILLER'S TROYER FAMILY

Around 1730s two brothers Michael and Andreas Troyer came to America on the same ship. The brothers were listed as members in the Northkill Amish congregation in Berks County. The family spread westward to Somerset County, Pennsylvania, and to Ohio and Indiana.

*Michael Troyer m. Magdalena Mast. Their children;

Michael, Andreas, Abraham, Joseph, *David.

*David Troyer children;

*Michael (DJH 9180), Christian, Barbara, Magdalena, Annie, Jacob, Henry, Mary, Joseph, Andrew, David, Frany.

*Michael Troyer m. Maria Reichenbach their children;

Abraham, *John, Elizabeth. Michael m. 2nd. children; Jacob, Daniel, Joseph.

*John Troyer m. Catherine Miller their children;

Samuel m. Magdalena Hochstetler b.1803 (7544)

John m. Magdalena Miller

Michael m. Barbara Miller 1810 (DJH 1930) dau. of Jonathan Miller

*Maria, 1806 m. Bishop Levi "Leff" Miller

*ancestors

A Bishop Levi "Leff" Miller Rocking Chair

SAILING SHIPS OF THE IMMIGRANTS

The sailing ship Phoenix *arrived at Philadelphia, Pa. on 15 Sept.1749, with some of our Miller ancestors, Hannes, Jacob, Christian, Peter Miller plus others. (young men and women were not signed.)*

The Bark Elizabeth Hall *of Dighton, arrived at Baltimore Md. 1 Sept. 1842, with some of the Bender family, (Mother)Elizabeth, Christian, Johannes, Jacob and Michael Bender*

CHRISTIAN "SCHMIDT" MILLER'S HAMMER

This hammer is said to have been made by Christian "Schmidt" Miller. Looking closely you will see the hammer is not perfectly shaped like a store bought one would be. But very well done for an 1800s blacksmith.

FERN STAND MADE BY ELIZABETH (WALTER) CHRISTNER.

Made with butcher knife, chisel and saw, now an interesting heirloom.

AUSBUND, CHRISTLIHE LIEDER 1785

I recently located an Ausbund that belonged to Christian "Schmidt" Miller and his son Jonathan Miller. The staff at Geauga Amish Historical Library, Middlefield, Ohio, were very kind and sent me a copy of the title page and a page with Christian "Schmidt" and Jonathan's handwriting, (in German script), I cannot verify the hand writing but the dates fall in the right time-line, as Christian (1763-1845) would have been 22 years old when he bought it and gave it to Jonathan (1789-1867) when he was 65 years old, and Jonathan was 43 years old.

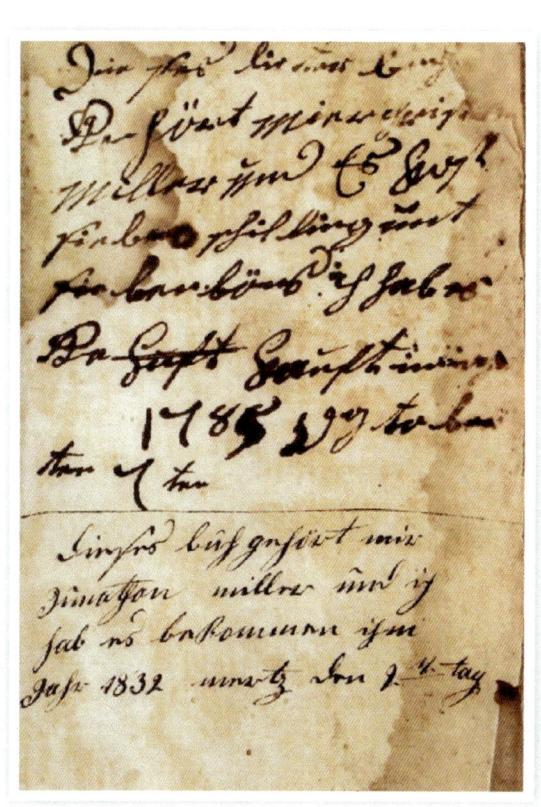

Dieses lieter buch
gehört mir Christ[ian]
Miller und es kost
sieben schilling unt
sieben böns. Ich hab es gekauft im [?]
1785 [October?] ten 7ten
Dieses buch gehört mir
Jonathan Miller und ich
habe es bekommen im
Jahr 1832 Mertz den 2tn tag

JONATHAN MILLER AND CHISTIAN "CHRISTAL" MILLER BARNS

The Jonathan Miller Farm, near Charm, Ohio. Holmes County.
Note small cemetery in upper right where Jonathan's wife Magdalena is buried.
Jonathan is buried in Lagrange County, Indiana.

THE HOMESTEAD OF CHRISTIAN "CHRISTAL" J. MILLER

This homestead of Christian "Christal" Miller, a mile south of Farmerstown, when State Road 557 was built about 1825, it went diagonally through the farm, between house and barn. In center of picture is "Chistal's" old house, later a new house and barn were built and the old house was torn down And a large house was built on this side of road, now there are two homesteads.

LEONARD'S LEVI "LEFF" MILLER CHAIR

This chair owned by Leonard Miller was made by "Leff" Levi Miller 1799\1884, Holmes County, Ohio. About 1850, they were assembled with wooden pegs and are as sturdy as when they were built. One wonders what kind of machinery he had at that time to build something like this, He had a lathe and other tools, powered by a horse going around a "sweep" The chair risers have marks on them that he used to align the spindles and mortise and tendon for the back pieces. It is said that when these chairs do come up for sale they bring very high prices. I am doing research to find out how many pieces he made, apparently he was very skilled and experienced, by looking at the workmanship. Three of his chairs are now in museums in Ohio.

LEVI " LEFF" MILLER CHILD'S CHAIR

A child's rocking chair owned by Henry Erb of Holmes County, Ohio, He says this chair has been used by generations of children in more ways then one, for instance they would turn it upside down and imagine it was a plow ? and scoot it across floor. Look at the spindles at top how they are worn down. Despite it's age and wear, and tear, it is still remarkably solid although the seat has been replaced how many times?

LEVI "LEFF" MILLER CHAIRS

Here are twelve of the chairs we found, there may be others that we don't know about. Three of these chairs are in Ohio museums, One in The Amish & Mennonite Heritage Center, one in The German Culture Museum, and one in The Colonial Homestead.

"LEFF JOHN" MILLER BARN

Built about 1862, the barn is 102 feet long, made from timber cut and milled on the farm, note heavy timbers and built in ladder inside. The barn is still in use today

EMMA'S BARN PAINTINGS

"Leff" John Miller Barn, painted on roof slate from the farm house.

*'Barn on Rocks' where we lived once when I was growing up.
These two barns were painted by Mrs. Mel (Emma) Hochstetler.
They live in Sarasota, Florida, Mel is my sister Alice's son*

CHRIS' FAMILY

*Chris and Alice, Robert and Carol Miller.
Justin and Rachel, Roxanne & Charlotte Minick*

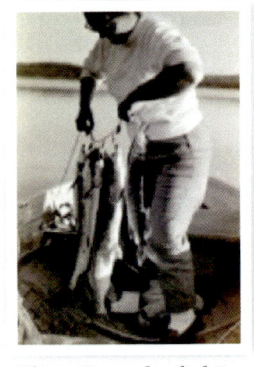

Alice Canada fishing

Chris, Alice and Robert

Rachel Miller/ Minick

Charlotte Minick

Chris' Family, Chris & Vera, Carol, Rachel, Roxanne. Justin. Front Robert, Charlotte

Roxanne Minick

Robert at work at Omnica

Chris and Alice on Canoe

In all thy ways acknowledge him

AND HE SHALL DIRECT THY PATHS

Prov. 3-6

My God shall supply
all your need according to his riches in
GLORY BY CHRIST JESUS.
Phil. 4-10

BOOK ELEVEN

Bishop "Leff John" Miller

THE MILLER-NISLEY SESQUICENTENNIAL FARM
By Floyd R. Miller Grandson of Manas DJH6157

Among the first Amish settlers in Indiana were Hochstetler descendents. In the spring of 1842 David H. Miller and his wife Anna (DJH 7805) the daughter of Peter Hochstetlers (DJH 7543), Moved to Elkhart County. They located about 5 miles east of Goshen in Clinton Township. On May 31, 1842 they bought a 160 acre farm from Isaac Smucker (DBH 9225) for $ 1200. Although David and Anna were in their late forties when they moved to Indiana they had no Children.

Within five years five of Anna's siblings also moved to Indiana.

It has been passed down through the generations that the original sod hut where the settlers first lived was located a little southeast of where the house is today. At this "hut" location, in the spring some daffodils would grow every year. The photo shows the daffodils and the spring close by where the first buildings were located. In the background you can see where the spring was located, (necessary water supply). Linda (Nisly) Miller, who lives there now, says the daffodils still come up every year.

Some years later when David and Anna were nearing the age of 60 they received some younger help on the farm when Preacher John L. Millers (widely known as "Leff John, i.e. son of Levi DBH 6749) of Ohio moved by covered wagon pulled by horses, onto the farm in 1862. John's wife Anna Hochstetler DJH 7699 was a niece to David's wife, Anna.

Conestoga Wagon

The story is passed down in the Miller family that on the evening of their arrival at the new home in the woods, a baby was born! Because the mother and the baby were getting more attention than the horses that had drawn the moving wagons, the horses got "lost" and ended up back in Ohio! We do know that a son, Jacob, was born on April 8, 1862 only three days before "Leff" John Miller bought the "back 40" which David's had owned earlier.

On April 30, 1862 John and Anna bought the 160 acre farm from David and Anna for $3000. The contract and life lease is recorded in book 25, pages 233 & 234 of the Elkhart County Recorder's records. The farm referred to is the southeast quarter of section eight in Clinton Township. In the contract and life lease David and Anna Miller reserve for their use as long as they live the house, barn, out buildings, garden, hay pasture and keep of two cows and one horse. They were also to have all the firewood and fruit they needed. Ten bushels of wheat and twenty-five bushels of corn were to be delivered to them annually for five years. "Leff Johns" were also to furnish David's with fresh milch and butter, do their laundry and bake bread for them for which David's were to furnish the flour.

According to the contract, David's were reserving their buildings, therefore

"Leff Johns" need to build another set of buildings for their home. Oral family history has it that "Leff John" built where the present buildings are. They built in the very center of the quarter section and there is supposedly a survey marker buried in the circle drive and the buildings are around that. He had the barn all laid out to be 100 feet long when he found out about someone else in the family was building one with that same length so John built his barn 102 feet long. The first summer must have been a busy one, with the need for buildings. During first their summer here on August 13, their daughter, Caty, died at age 1 year 8 months and 27 days. She was buried on the farm on the knoll south east of the buildings. This may have been the start of what is now known as the Nisley Cemetery, But with all the fieldstone markers in that cemetery some of those may have been earlier.

John L. "Leff John" 1826-1890 m. Anna Hochstetler 1828- 1913
Their chidren;
Joni.(Jonathan) J.L. 1845-1902 m. Lydia Garver Goshen, In.
Joseph J.L. 1848-1922 m. Susanna Keim White Cloud, Mi.
Mary 1850-1870 m. Jacob Nisley
Levi J. 1851-1852
Seth J. 1853-1932 m. Mary Miller Morton, Il.
*John J.L. 1855-1938 m. Mary Christner Goshen,In.
Magdalena 1857-1944 m. Joseph E. Chupp Goshen,In.
Simon J. 1859-1929 m. Lydia Schrock Topeka, In.
Catherine 1860-1862
Jacob J.L. 1862-1945 m. Lucinda J. Chupp Shipshewana,In.
David J. 1864-1943 m. Catherine Bontrager Shipshewana,In.
Daniel b. 1866
Anna b. 1868-1869
Susanna J. 1870-1949 m. Samuel S. Eash Topeka, In.
Noah J.L. 1872-1943 m. Mattie Miller Sturgis, Mi.

In 1866 "Leff John" was ordained Bishop of the Clinton Amish congregation.

At this time his father-in-law, Joseph F. "Sep" Hochstetler (DJH7602) was the deacon in this district and "Leff John's" brother-in-law, David J. Hochstetler (DJH6145) was one of the ministers.

(David Hochstetler was married to Joseph P. and Magdalena Hochstetler's daughter Magdalena- a younger sister to Anna who was married to "Leff John". David and Magdalena were married in 1861 and lived on the family farm, also in Clinton Township with her parents. In that decade they also built a large barn –40X100- feet which may have been the source of friendly rivalry between these Amish preacher brothers-in-law! I know this latter "smaller" barn well, as my parents bought our historic family farm in 1948 and I spent my teenage years there.) *(Daniel Hochstetler)*

In 1885 "Leff John"'s son (referred to as young John, (DJH7718) was ordained to the ministry. On March 21, 1890 "Leff John" passed away at age 63. at this time Anna's father, Joseph P. Hochstetler, was still living but another one had been ordained to take care of the deacon responsibilities.

A year later the undivided two thirds of the farm was sold by Joni as administrator to his brother John who after two years sold his share to Joni. In1898 the widow sold her share to Joni as well. Joni and his family lived on the farm till his death on February 6, 1902 at age 56 years.

After Joni's death his widow, Lydia, and their children sold the farm to daughter Anna and her husband Jonas Nisley (DJH 3254). At this time "Leff John's widow Anna was still living with them. So when Jonas and Anna bought the farm, both her mother and grandmother were living with them on the farm. On March 18, 1930, Jonas Nisley passed away at 55 leaving a widow and two sons on the farm. Three years later son Perry got married and bought the farm from his mother the following year. Perry and Betsy then raised their family here. Later they lived across the road (C.R. 34) a few years before returning to the Dowdy house on the home place, Perry's sold the farm to son Glen and wife Leanna in1978, who currently live on this historic farm. (2008)

This Miller-Nisley farm has now been in the same family since the David H.

Millers bought it on May 31, 1842, for a total of 165 years (2008). I am not aware of any Amish farm in the community that has been by the same family longer than this one. (Steven and Linda (Nisley) Miller live on the farm now. (2015) The lane to the Nisley farm goes off to the north of C.R.34 while the cemetery is accessible from C.R. 35, There is a large rock by the entrance engraved NISLEY CEMETERY.

The "Leff John" farm located 4 ½ miles east of Goshen, Indiana. The Barn drawn by Velma Peck (Leff John, John J., Jacob, Clarence, Velma)

HOCHSTETLER-MILLER CONNECTION

1. Jacob Hochstetler m. Miss Lorenz-- 6526 Joseph Hochstetler m. Anna Blank-- 7543 Peter Hochstetler m. Madlena Yoder-- 7602 Joseph Hochstetler m. Magdalena Eash-- Anna Hochstetler DJH 7699 married "Leff John" Miller, son of Bishop "Leff" Miller.

THE JACOB HOCHSTETLER FAMILY AND THE INDIANS

Jacob Hochstetler, his wife and two children came from Europe to America in 1736. In 1736, with four children at home still at home, they were living in the Northkill settlement. Close to what is now Harrisburg, Pennsylvania. The French and Indian War was raging, and the Indians attacked many settlers in their homes on the frontier.

On the evening of September 19, 1757, the Hochstetler family had an apple snitzing for the young people. After the guests had left at a late hour, the family went to bed. The dog awakened son Jacob, who went to the door and looked outside to see what was wrong. When he received a shot in the leg, he immediately knew they were being attacked by Indians and he quickly closed and bolted the door.

The family in the cabin consisted of Jacob and his wife, Jacob Jr., a daughter (name unknown), Joseph and Christian. Joseph and Christian were good marksman as was their father. And the sons urged their father to allow them to defend themselves from the Indians which were only about eight or ten in number. But Jacob steadfastly refused, saying it was not right to take another's life even to save his own.

The Indians set fire to the Hochstetler cabin and the family went to the cellar. They checked the flames above them somewhat by sprinkling cider on the ceiling. This was not sufficient and as the flames spread the family decided to escape through a small window. Daylight was approaching, and the Indians were beginning to leave the scene of destruction. However a young warrior (Tom Lions) who had lingered in the orchard to eat peaches, saw them and called to his comrades.

Quickly the family was surrounded, the daughter and son Jacob were tomahawked and scalped. The mother was stabbed in the heart with a knife and scalped. According to tradition, some years earlier a band of Indians had asked Mrs. Hochstetler for some food and she had refused them. It seems these Indians now had a grudge against her because they used this method of killing her. The father Jacob and Christian were taken as prisoners and Joseph who had temporarily escaped, was also captured.

John, a married son who lived on the adjoining farm, saw that the Indians were his ust as the Indians were finishing their bloody work. A daughter

Barbara was likely married at this time, at she was not at her parents home.

The three prisoners were taken to the Indian village. They escaped having to run the gauntlet by making a present of some of their peaches to the chief. The three were separated. But before they parted, Jacob instructed his sons to remember at least their names and the Lord's Prayer.

Jacob tried to appear contended, but he watched for a chance to escape, this opportunity came about three years after his capture. The able bodied men left camp to raid the white settlements. Leaving Jacob to hunt game for those at home. Jacob was given a limited amount of ammunition, but he managed to save some of it.

The Indians would not tell him where he was in relation to the white settlements, but one day he managed to get a little idea of his whereabouts from seeing an old Indian draw a crude map in the dirt. He decided the time for escape had come.

Jacob traveled mostly at night, when he finally reached what he believed was one of the head waters of the Susquehana, he built a crude raft and began drifting downstream. He did very little shooting because he was afraid of being heard by the Indians. He had hardly anything to eat and became weak with hunger. When he finally came to Fort Harris, the site of present-day Harrisburg. He was unable to stand and no one observed his attempts to draw notice from those in the fort, however below the fort, a man watering his horse saw strange object and reported it. The commander of the fort looked through his spy glass and saw a man on a raft. So Jacob was rescued and was once again with his people.

Jacob's sons Joseph and Christian were returned with other white captives sometime after the close of the French and Indian War and possibly 5 to 7 years after their capture. The boys had been adopted by the Indians and had been treated as kindly as if they had been one of the Indian's own race. It was with reluctance that the boys left their Indian friends, and they occasionally returned to visit them after their release.

Details for this story was taken from <u>Descendants of Barbara Hochstetler</u>

(1938), by Harvey Hostetler.

The first Hochstetler book by Harvey Hochstetler was printed in 1912 which established a data base for most of the Amish in America can relate to including the "Miller" connection. There are probably more Millers in the Untied States than there are Hochstetlers, but we never had a data base established and now it is too late and over whelming with the numbers.

My parents (Jacob J. and Ada Miller are in the 1912 Hochstetler book, as are my two oldest brothers Oba and Clarence. As well as many of my other relatives.

BOOK TWELVE

John J.L. and Mary Miller

*John J.L. Miller (My Grandpa)was born March 15, 1855 in Holmes County, Ohio. Son of John "Leff John" and Anna Miller, And came with his parents to Indiana in a covered wagon when he was six years old. Mary Christner was born in Elkhart County, Indiana, April 6, 1858 daughter of Jacob and Elizabeth (Walter) Christner. They were married near Kalona, Iowa December 25,1877. to them were born 10 children. *John was ordained as a Minister of the Amish church in 1885 and served in that capacity until his memory failed him and was unable to continue.

This cradle, sold at Uncle Oba's farm sale is said to have rocked all of John and Mary's children when they were infants.

John J.L. Miller 1855- 1938 m. Mary Christner 1858- 1923

Their children;

David J.C. 1878- 1948 m. Elizabeth Jantzi Elkhart Co. In.

Levi J. 1881- 1959 m. Mary Schrock Elkhart, Co.In.

John J.C. 1883- 1965 m. Mary Hochstetler Goshen, In.

*Jacob J.C. 1885- 1959 m. Ada C. Miller Elkhart, Co. In.

Harvey J. 1887- 1946 m. Fannie Helmuth Amboy, In.

Mary J. 1889- 1926 m. Samuel J. Kauffman Middlebury, In.

Obadiah J. 1891- 1975   m. Elmina C. Miller,   m. 2nd to Anna M. Yoder Amboy, In.

Elizabeth J. 1893- 1918 m. Jacob V. Lambright Elkhart, Co. In.

Mattie J. 1896- m. Andrew J. Yoder Lagrange Co. In.

Fannie J. 1898- 1962 m. Menno Kuhns Nappanee, In.

The 1892 plat book for Clinton Township, Elkhart Co., Indiana shows a total of 289 acres belonging to *John J. miller, part of which was the "Leff John" home place where he grew up.

They moved to Reno County, Kansas in 1895 Then to McPherson County, Kansas.

My dad, Jake told many stories of living in Kansas near a railroad track, but unfortunately I don't remember all of the stories he told and I don't know where they lived at exactly.(I will write some of those stories at another place " Dogs, Wolves and Coyotes") Apparently this was an exciting growing up place for dad as this was when the west was still wild and wooly, The prairies were being plowed, fences were going up, railroads were just new and cowboys and coyotes were a common site. Geronimo was still running loose, cattle drives and cow towns were in their heyday. As young boys I think they thoroughly enjoyed life in the old west.

Doing research I found that the Chicago, Rock Island and Pacific "CR&P" railroad ran through Kansas in 1852. it passed through McPherson and Reno county, with stops at Inman, where many Amish settled, then on thru Hutchinson, Partridge and on south west. So if they lived near Partridge they would have been near the railroad.

The train engines of that era were the 4-4-0 Type American style and most

4-4-0 American Engine, drawn by Velma Peck

of them were wood burners as was the most available and cheapest fuel at that time. The coaches were not very comfortable, as with wooden seats and no cushions, but it was better than wagon trains.

And also the Santa Fe wagon trail ran through the Amish settlement. The Santa Fe Trail was a major route for migration from the eastern states to the new southwest frontier, the trail was at places wide enough that four or five wagons could drive side by side, what a sight that was with covered wagons as far as you could see. But when the rail roads came through the trail was abandoned all that is left are some wagon tracks that you can still see today.

In 1898 they (John's) moved to Jackson County, Minnesota then later to Nobles County,

Minnesota House by Velma

Minnesota, where there was an Amish settlement of about a dozen families, some were from Indiana, Illinois, Michigan and Ontario, Canada.

This was prairie land the prices were cheap (as low as $14.00 acre) this was raw prairie land, never having been plowed, After breaking the tough sod, the farmers sowed the large fields in flax. The next year they sowed wheat or oats and the third year they planted corn. Most of their plowing was done in the fall until the ground froze which was often in the middle of November. In the spring they sowed their wheat in March or April, Spring wheat rather than winter wheat was grown.

When the grain was ripe, it was cut, tied and stacked. During the following months it was threshed. Amish youths came from various states to work in the Minnesota grain harvest.

In 1898 John J. Miller purchased a threshing machine which David D. Schlabach operated with his steam engine. Threshing was a long drawn out activity for the farmers. In October 1895 the threshing was only half finished. It was not unusual to continue threshing into the winter months with the men working as the weather permitted. But sometimes it could not be finished before deep snow and cold weather came.

Some of the grain was cleaned and shipped as seed. One year *John J. Miller and David D. Schlabach shipped fifty-five bushels of wheat seed to Wright County, Iowa.

Their son David J.C. Miller married Lizzie Jantzi there on on August 5, 1900. Their son David was born April 28, 1902, died there (in Minnesota) of measles on May 2, 1903.

With harsh winters and other disappointments of every day life plus disagreements in church matters, members began leaving the area, some going back to Michigan or

Indiana and other states. The *John J. Miller family moved back to Elkhart County, Indiana in 1905.

I don't know where grandpa's lived or what he did at that time, but around 1915 four of his sons (David, John Jr., Obbie, Jacob) moved to West Branch,

Oscoda County, Michigan. *John J. also bought land there but never lived there. The venture there was short lived as the four brothers moved back to Indiana about 1919.

In about 1920 *John and Mary moved in with their son Levi, for a number of years. When Grandma (Mary) became very ill, she desired more solitude and they moved away for a short time till she died March 2,1923. Grandpa then moved back to Levi's where he remained about five years, then transferred to son *Jake's where he was for six years. He returned to have his home with Levi's in1934 where he died in 1938.

He was a carpenter. In his retirement years he still would make (wagon) hay racks to sell. My Dad had a hayrack that fitted on different running gears, I wonder if Grandpa made that? He had a small shop where he liked to spend time in making little things (small benches, toys, etc).

He was a fisherman and a number of grandchildren remember going fishing with him. He also had a fish route – peddling them. He would get them in large wooden boxes (by train) packed in ice and go from house to house. As a little girl, I (Levi's Gertie) often went with him and delighted (as also did my brothers) in jingling his little bell announcing our arrival as we drove up with his horse "Tops" in his spring wagon. He was a bee keeper, he had honey and bee supplies to sell to his neighbors and kin for many years. (My sister Nancy related when she was little and he was tending his bees, she would set on the ground close by and play wearing only a sun dress and the bees never bothered her). As his memory failed and he was unable to work, he grew very restless and did a lot of walking, always on the move in and out. (Bro. Olen told me he remembered grandpa walking out to the woods and back with his hands behind his back and smoking a pipe with smoke going back over his hat much like a steam engine).

On September 8,1938 he fell and broke his hip which resulted in his death on October 3,1938. I (Chris) remember as a young boy about 7 or 8 years old standing by his sick bed, he put his arms around me and talked for a long time,

I don't know what he talked about, but now I sure wish I knew what he said, I'm sure they were words of wisdom and inspiration. During those last weeks, he would often be heard singing parts of old German hymns, expressing his desire to go home with the Lord.

I remember him as a kind and gentle man and my Grandpa.

It's a shame that so many **OF US WHO CAN TELL THE WORLD HOW** to solve its problems haven't been able to solve our own.

BOOK THIRTEEN

The Christner Family Branch

Switzerland

*Christian Christener m. Barbara Burkhard, Oberdiesbach, Switzerland

Switzerland

*Christian Christner b. 1721, Switzerland moved to Alsace, France

France

*Christian Christner b. 1750, Switzerland d. Windstein, Alsace, France m. Margarete Stucki

Germany

*Peter Christner b. ca. 1766 m. Magdalena Guth moved from Windstein to Bavaria, Germany

Canada

*Christian Christner b. 1799 d. 1836 Ontario, Canada. m. Elizabeth b. 1796\1862 Wilmot Twp. Ontario, Canada.

Indiana\Illinois\Indiana

*Jacob Christner 1828 – 1893 m. Elizabeth Bontrager 1833 - 1851

m. 2nd *Elizabeth Walter 1826 – 1884 m. 3rd Anna (Albrecht) Kennel 1838 – 1910

(Jacob and Elizabeth Walter), Their children;

Susanna, Elizabeth, Samuel, Frany, *Mary, Barbara, Joseph, Jacob, David.

Indiana\Kansas\Minnesota\Indiana

Rev. *John J.L. Miller 1855\1938

His parents, Bishop "Leff" John L. Miller 1826\1890 Anna Hochstetler 1828\1913

*Mary Christner 1858\1923

Their children;

David, Levi, John, *Jacob, Harvey, Mary, Oba, Lizzie, Mattie, Fannie

Indiana\Michigan\Indiana

*Jacob J.C. Miller 1885\1959

Ada C. Miller 1886\1963

Her parents, Christian C. Miller 1856\1923 *Mary Bender 1856\1899

Their children;

Barbara Yoder Oba, Clarence, Barbara, Mattie, Alice, Olen, Saloma, Nancy, Lester, Chris

THE CHRISTNER FAMILY STORY

The first reference to an Anabaptist Christner, or Christen was Christen Christen, who came from Bembrunnen near Langnau in the Emmental and was put in the Zuchthaus (prison) in Bern for his Anabaptist beliefs. Christen is mentioned in the *Martyrs Mirror* as a prisoner in 1659 with other Anabaptists. A document of 1661 mentions some of the same Anabaptists, with names like Jacob Gut (Good), Jacob Schlabach, Hans Zaugg, and Hans Wenger as fellow prisoners. They were given three choices: to return to the Reformed Church, to be banished to the galleys, or to be executed. Luckily, some managed to escape, and others were hidden from the authorities in their villages.

Diessbach between Langnau and Thun became a gathering place for Anabapists back in remote valleys. It also was the rallying place for followers of Jacob Amman who preached a more conservative Anabaptist way of life.

A group of these received the order to be exiled to Alsace, and were warned never to return. Before that in Switzerland, they had to remain in very out-of-the-way places, even having to worship in caves. They were not allowed to build church buildings, and perhaps this memory has led to the followers of Amman to worship in homes to the present time.

Christen Christener (an original spelling) and Barbara Burkhart, were probably not the first Anabaptists with the surname of Christner.

As early as 1660 a Christian Christner was taken into prison at Bern, Switzerland, along with ten other prisoners. He had been expelled from his homeland and placed aboard a boat on the Aare and Rhine Rivers. The surname Christner is not as common as some other names in the Amish and Mennonite Church. Also records and stories at the time were not kept as well as they were later on and some were lost or destroyed, So tracing family history is sketchy at best and leaves many empty spaces which we cannot fill and will not attempt because it would be pure speculation, we will leave that up to the novelists.

Peter Christner was born near Windstein, Alsace about 1775 his wife was Magdalena Guth. They moved to Bavaria in 1809, He was the son of Christian and Margarete Stucki our possible ancestors.

Beautiful Bavaria! a southeastern state within Germany lies mostly on a plateau surrounded by mountains which rise to their highest on Mt. Zwgspitz at 9,721 feet. Today, Bavaria is a tourist center with people from around the world coming to visit the beautiful mountain and lake regions.

While it is now remembered in nostalgic and pleasant ways, it has not always been so. Within the memories of our Anabaptist history, one remembers that it was here that some of the greatest persecution was exacted. Bavaria was a place both to flee to and to flee from, depending on the ruling magistrates, All laws regarding them were simple. All Anabaptists will be punished with death, those who recant will be beheaded, those who remain stubborn will be burned. Henceforth none of them will escape execution even if they recant. *The Martyrs Mirror* records hundreds of deaths. We have reason to believe our ancestors Christian and Elizabeth sojourned here for a time since at least one of their children, was born here.

Beautiful Bavaria

Napoleon occupied and later decreed Bavaria a kingdom declaring that no religious group could be relieved of military duty. It was at this time that the second big migration of Amish and Mennonites from Bavaria and Alsace Lorraine began to leave for Ontario, Canada and the United States of America.

Evidently our ancestor Christian Christner's first children were born in Germany then Bavaria, back to Germany and Wurtenberg, Where *Jacob was born, then on to Alsace-Lorraine France. From other stories we find that the Anabaptists were being harassed and they kept moving from place to place trying to find peace for them and their families and could not find a safe place and they eventually moved to the "New World". They found their way northwest to the Netherlands and passage across the Atlantic and the New World.

Christian and his wife Elizabeth, with their first five children, (Lizzie, Magdalena, Peter, Catherine and *Jacob, came to America in 1828. Their last three children (Daniel, Mary and Christian Jr. were born in Canada, in the Baden Ontario area.

On the ship from Europe, Jacob, a baby of a few months, was believed to be dead by the one on the ship who looks after such things, as when people died at sea. He wanted to throw Jacob out into the sea, the reason being that there was a superstition that the whales (sharks) would smell the dead and molest the ship. Elizabeth cried so for her baby that they let her keep him. She prayed and prayed for her baby until he finally moved a finger, and they knew he was alive. (As a young boy I heard my mother relate this story many times and I dismissed it as a tale, not realizing that this actually happened, and to my own Great-grand father! Another miracle is that hardly any children under one year old ever survived the voyage to America at that time.

In Baden, Canada as Jacob was eight years old, when he and his father, Christian were out in the woods cutting down trees and clearing land. Jacob cut a tree down that fell against another large tree. Christian cut the larger tree down, and in some way got killed by the falling tree in about 1836.

The family was believed to have a small home by that time, But by the time everything was settled, there was nothing left. The mother, Elizabeth, was put out and forced to work for others to make a living, and the children were placed in homes to work for their own needs. The children who were over sixteen were free to go for themselves.

Later the mother, Elizabeth, walked down to Indiana, probably in the 1840s, with some of the children, They crossed the Great Lakes near Detroit, Michigan, with a covered wagon and while crossing, the log raft they had built started to come apart but with a little ingenuity and endurance, they were finally able to land safely. After which some time was spent in town (Detroit). Tradition has it that Elizabeth along with her children walked to the new land, leaving the wagon to mainly carry belongings, provisions etc. The able bodied were expected to walk.

When they arrived at Goshen, Indiana, Jacob began clearing land, cutting trees with an axe and built a log cabin and farm buildings east of Goshen, Indiana along what is now C.R.34.

After living in Elkhart County, Indiana for several years Elizabeth walked to Wayland, Iowa, and lived with her daughters. She is buried just north of Wayland in the Sommers Cemetery next to Her daughter, Magdalena.

Jacob Christner was first married to, Elizabeth Borntrager on Nov. 14, 1850 She died of childbirth Apr. 1, 1851. Aged 17 Years and 10 Months. She was daughter of Joseph and Barbara (Yoder) Borntrager.

Jacob Married second time to Elizabeth Walter daughter of Samuel and Frances (Borntrager) Walter Aug. 14, 1851. They had nine children.

The homestead was on the tract of land remembered as the Sol Schrock farm on County Road # 34, or Fish Lake Road, East of Goshen, Indiana. This site Jacob had chosen consisted of at least 130 acres of land and was homesteaded by Jacob and his wife Elizabeth at about the time they were married in 1850.

Much of the land needed clearing, and the family stories tell of Jacob clearing the land and constructing the first buildings on this site.

One of the original barns built by Jacob is still in use and in good repair, This barn was built the same year Jacob's fifth daughter was born in 1860.

Much of this lands topography, when cleared, was gently rolling and good farming soil once the stumps were removed and the soil prepared.

A letter from Olen J. Miller, Great-grandson of Jacob Christner, wrote- We have in our possession the axe (head) which was used by

Jacob Christner to "clear the way" when they came from Baden, Ontario, Canada to Elkhart, County, Indiana around 1848. At that time there were more forests and swamps throughout this area, and they often had to chop down trees to make a path to get through. They would also at times make use of the chopped down trees and lay them across the swamps to make a road way for their wagons.

Jacob Christner later gave this axe (head) to his son Jacob. Around the year 1930 this Jacob Chrisrner, who was living in Iowa, came to Indiana on a visit, bringing this axe with him and gave it to my father Jacob Miller, since his name was Jacob and said he should have it.

After my father passed away, Mother came and gave the axe to me and said, "Father always said you should you should have the axe when he isn't here any more". And it is still in the Olen Miller family, It would be over 150 years old by now (2014)

When Jacob decided to move to Shelby County, Illinois (about 1872) he sold the farm to Simeon Miller who is the great-great-grandfather to the present owner Rod Bontrager.

Simeon was also the father-in-law of Barbara Christner, one of Jacob's daughters. Barbara's husband was Joas Miller the son of Simeon Miller and, while the farm did not stay in one line or family name, it nevertheless was transferred to close relatives for the generations since Jacob decide to sell it.

The house on this farm burned to the ground in1889 after which the old foundation, with additional foundation, was used to construct another house. In the stairwell leading to the attic, there are a number of things written on walls on either side I want to do more research and obtain photos if I can to learn what is written there.

Jacob Christner spent his married life east of Goshen, Indiana, Except for the time in Shelby County, Illinois, approximately 1872-1882. The account of the Shelby County group's intention to form an extra, low order church seems to support the information that Jacob moved to Shelby County because he was dissatisfied with many things, He strongly opposed the use of windmills, saying he didn't believe it was right to expect God to pump your well water.

About eleven years after it began or 1883, the Shelby County group disbanded. Several sources indicate that there were church problems which could not be resolved.

Our ancestor, Jacob Christner probably better off than most, returned to Indiana in1882 and again settled at Elkhart, County in Clinton Township.

Jacob and Elizabeth (Walter) Christner had nine children all born in Indiana.

Elizabeth (Walter) Christner died Dec. 14,1884 at the age of 58 yrs. It is thought that she is the Grandmother who made some furniture. She used three tools, a butcher knife, a chisel, and (a saw?. Elizabeth's daughter Bevy wanted each of her sons to get a piece of this furniture. Obie had the couch. Sam had the rocking chair, and Ezra had the plant stand (now it belongs to his son Harvey's family).

Jacob remarried the third time to Anna (Albrecht) Kennel on Mar. 18, 1886 (married 7 years) born Mar 19, 1838 Died Jun. 23, 1910

Anna migrated from Canada to Goshen, Indiana (Elkhart County). After Jacob's death she moved back to Canada.

Clinton Union Cemetery East of Goshen, In. on C.R. 36, west of C.R 35.

Nine gravestones in a row, all members of the Jacob Christner family

*Jacob Christner b. Alsace, France Apr. 11, 1828. d. Elkhart County Indiana Mar. 13, 1893 age 64 yrs. m. Elizabeth Walter b. Feb. 1, 1826, Holmes County, Ohio d. Dec.14 1894 Elkhart County, Indiana, age 68.

Chn.

Susanna 6-10-1852 d. 11-23-1881 m.1870 to Christian J. Troyer 1847-1917 Goshen, Ind.

Elizabeth 7-19-1853 d. 3-18-1935 m. 1875 to Jerry D. Yoder 1852-____ Haven, Kansas

Samuel 11-12-1854 d. 8-20-1937 m.1859 to Elizabeth Mast 1859-1916 Huron County Michigan

Franey 4-23-1856 d. 5-16-1862 m. 1881 to Simon Schlabach 1851- ___ Goshen, In.

*Mary 4-6-1858 d. 3-22- 1923 m.12-25-1877 to *John J. L. Miller 1855-1938 Goshen, Indiana

Barbara 4-6-1860 d. 4-25-1921 m. Joas S. Miller 1857-1890 Goshen, Ind.

Joseph 6-18-1861 d. 5-16-1862

Jacob Jr. 9-23-1862 d. 2-6-1939 m. Fanny Yoder 1861-1929 Kalona, Iowa

David J. 3-10-1865 d. 4-10-1941 m. Lovina D. Raber 1865-1938 Shelbyville, Ill.

THE WALTER FAMILY

*Samuel Walter 1793- 1867 m. Veronica "Frany" Borntrager 1796- 1881("Frany" was the sixth child of the first John Borntreger who lived near Meyersdale, in Somerset Co.,Pa.)

Samuel and Frances lived near Barr's Mills, Ohio, they had a family of 13 children. In 1848

The Samuel Walter Family Bible (Now in Amish library % Floyd Miller)

they moved from Ohio to Elkhart County Ind., five miles southeast of Goshen, In. (S. E. corner of Cr. 36 and 35).

**Samuel and **Frances are buried in Clinton Union Cemetery.

Their children;

John (infant)

Jacob, He married and lived five miles east of Goshen, where he worked at the blacksmith trade, later he moved west.

Samuel, He remained single and went to California when gold was first discovered there, later went to Oregon.

Mary, married Samuel Blough in Lagrange County,In. Mennonite church. Chn. Amos, John, Rebecca, David, Frances.

Barbara, married Jonas Edwards, Goshen, In.

*Elizabeth, Married Jacob Christner (see ancestor Jacob Christner)

Joseph, He came with his parents to Elkhart Co. In. He later went west.

David, He came with his parents to Elkhart Co. In. and later went to Iowa.

**Amos, Married Adaline Hani, Shipshewana, In. They took care of his mother Frany until she died. Chn. Edwin, (infant), Minnie Elizabeth, Amos, Nicholas.

**Frany, Married **Nathan Inbody, Topeka, In. chn. Eveline, William, Nettie, Frank, Sarah.

Rebecca, married Jonthan Stutzman, Inman, Kansas. Chn. Amanda, Samuel, William, Frank, Viola.

**Susanna, married **John Dunlap, Goshen, In. chn. **Ruben, Samuel, John, **David, d.

Gertrude, married Lewis Scalf, Goshen, In.

** These are buried in Clinton Union Cemetery.

BOOK FOURTEEN

Jacob J. C. Miller (Jake)

Jacob J.C. Miller (Jake) was born November 11, 1885 while his parents, John J.L. and Mary (Christner) Miller lived on the "Leff John" farm, this would have been Dad's Grandfather, east of Goshen, Indiana. In 1885 Grover Cleveland was President, the first skyscraper was built in Chicago, Ill. (10 floors), The United States flag had 38 stars and the first practical bicycle was made.

His parents, moved to Kansas in 1895, Jake would have been 10 yrs. old. They first moved to Reno County then later moved to McPherson County,

American 4-4-0 by Velma

near Wichita. I don't know how they, (Grandpa's) moved but Dad talked about traveling by train. In 1892 about 10 families moved from Middlebury, with a coach and baggage car escorted to Kansas by a land agent. They settled close to the railroad as Dad talked of an incident that happened when they lived there. One Sunday afternoon, the boys not having much excitement, were walking along the railroad, when along came a track inspector on a small hand car, he stopped and talked to the boys and informed them he was going to Wichita. They asked if he was coming back and said oh sure and they could ride along if they pumped the handles to make it go. Alright that sounds like fun so down the track they went with the inspector sitting with crossed legs over the side and he lit up his pipe and enjoyed the ride while the boys did all the work. When they came to Wichita he took the car off the track and pushed it into a small metal shed and locked the door! Wait, aren't you going back? Oh yes, tomorrow! TOMORROW? Oh yes, you didn't ask me when, he replied. Now that left the boys a long way from home and they knew they had to be home to do the evening chores, so they had hustle as fast as they could running down the railroad tracks, which is not very easy if you ever tried it, it is not very easy to run down the railroad tracks no matter on the rails, the ties, or on the ballast rocks. I guess they made it home in time but they learned a hard lesson.

One time there was a carnival going on in Wichita and some of the boys decided to go and spend some time there on a Saturday afternoon and evening. When it was time to go home they decided to ride a freight train home. They asked the brakeman and he gave them permission to ride one of the freight cars. They knew there was a small grade near their farm, where the trains had to pull hard and slowed way down to go over the hill, they figured they could jump off with out any trouble, only thing was when they got close to the hill they were going at a high rate of speed and the boys knew if they didn't get off it would be a long way back. So they decided to chance it anyways and jumped.

Landing in the bushes, rocks and what not, they were scraped and cut all over their bodies when they landed.

At that time it was common to see cowboys, on a cowpony, with their gear and a six-shooter in a holster, Coyotes running around that had not learned the fear of man and many people had the fear of the Coyote, a kin to the Wolf and not to be trusted.

When Dad went to school (He said second grade) they had to walk ¼ mile to the corner then 1 mile to the school house. One day as they were walking home, Dad, and an older brother and a younger sister, as they came past a straw stack a little ways from the road, there were three coyotes sitting on top and watching them, Dad said they were licking their chops!! They dared not run or look scared so they held there little sister between them and told her not to make any noise, but she kept sniffling and was sure they were going to be eaten by the Coyotes! They kept walking until they came to the corner.

Then they picked up their sister by her arms and began running for dear life, Dad said they were running so fast that their heels were going past their ears and the sister was running just as fast, only about a foot off the ground, with her legs going and sobbing all the way home! Dad told this story so many times I memorized it.

Dad said that was the last they went to school there, although they went to school later in Minnesota.

Dad also used to tell how he and his friends spent Sunday afternoons catching gophers (ground squirrels), they would take a string and made a lasso and placed it around the gopher hole then sit back and wait, the gophers always made their holes at an angle and they would always emerge at the slope side of the hole, so they always sat on the opposite side, so when the gopher emerged facing the other way, they jerked the string and caught the gopher!! As gophers were very plentiful it made for an exciting afternoon for the boys.

Dad told me many other stories about Kansas, I think it was an exciting time for the boys at that time.

One threshing season Dad was given the job as fireman for the steam engine that powered the threshing machine. He thought that was quite an honor for a young boy, especially when they told that when he had the engine fire going, he could sit in the shade and relax. So he worked real hard, he had to carry the straw from the back of the thrashing machine to the back of the engine, as that is what they used for fuel because it was plentiful.

He got the fire going real good and had the steam up and everything was running smoothly, so he decided to get a drink of water from the jug and sit in the shade for awhile, about that time someone yelled "Hey Jake your fire went out" then he had to start all over again, that's when he found out he had the hardest job of anyone on the crew. Dad had a brother John and when they lived in Kansas, one day when they were working in their farm blacksmith shop, John as a teenager threw a hammer toward a drawer of tools and when it landed a chip of iron flew off and hit John in the eye.

Not being able to remove the chip and the closest primitive medical facility being a long ways off by horse and buggy, the time was in the 1890s and the closest town was Hutchinson, Kansas. They probably used home remedies but nothing helped and it got infected and became extremely painful and He kept begging to go to the Doctor, so his Dad said Ok if it's not better by morning, we'll take you to the Doctor.

The next morning John was in such misery that he got up and done his chores, got the horse hitched to the buggy and was ready to go before the rest of the family had breakfast.

When they did get to the Doctor, he said the eye could not be saved and they had to remove it. Dad always said John must have had a real high fever as he was not the same afterwards, He had problems all the rest of his life.

Coyote

One story Dad often told about John was that

when one of the other boys were working in the fields with the horses and spike harrow, a couple of coyotes started following him, and their Dad, thinking they might be wolves, had told them if that happens, to unhitch the horses and ride one of the horses back home.

During dinner John started talking how disgusted he gets at the coyotes and said he had a plan to fix those coyotes, but wouldn't tell them what it was. After dinner they couldn't find him anywhere. Finally they found him out behind the farm shop, he had two straight files and was sharpening the pointed ends. When the boys almost beat the stuffing out of him, He finally told them what he was going to do with the files, He said he was going to put the files in his pockets, one on each side and while he was harrowing this afternoon and the coyotes came around He would take files with the sharp points and slash at the coyotes when they came close enough and nothing they could say or do would keep him out of the field that afternoon.

So he hitched up a team of horses, and took them out to the field and hitched them to the harrow and went to work as usual. Along about middle of the afternoon when he had all but forgotten about the coyotes and his files, plodding along behind the harrow and the slow moving horses in the dust and hot sun. When all at once he heard a neighbor yelling like an Indian at him and when John looked around there were three coyotes close enough he could have reached them with the buggy whip.

Well John took one look at them and never thought of his files or anything else, He just took off across the harrow and without unhitching the horses, he jumped on one of them and went for the house as fast as he could, with the harrow bouncing along behind. Needless to say he was reminded many times of his bravery.

In 1898 Dad's parents moved to Jackson County, Minnesota, then later to Nobles County. The country there was more harsh than Kansas, but the land was cheap! it was mostly prairie with very little timber for building and for fire wood. The growing season was shorter and the winters colder and snowier. They did raise a lot of wheat and His Dad owned a threshing machine, so Dad probably got more experience with engines and threshers.

He also got more schooling, I have in my possession a small pamphlet that Dad got when he went to school in Nobles County, Minn. In it lists all the pupils attending the year 1899-1900. Jacob and five of his siblings, Lizzie, Harvey, John, Mary and

Obbie.

On the 2nd page there was a place for photo of the pupil and you can see it was torn off because they were not allowed to have photos, Oh how I wish I could find that photograph!! He would have been fifteen years old, what did he look like then ??

Eventually most of the Amish left that area, some moved back to Indiana and Michigan and other states.

Dad's parents moved back to Elkhart, County, Indiana in 1905.

ADA C. MILLER MY MOM (MEM)

Jacob J.C. Miller married Ada C. Miller on January 2nd 1908, Ada was the daughter of Christian C. and Mary (Bender) Miller of Lagrange County, Indiana, Ada grew up at what is now known as 3640W 350S Topeka, Indiana Where a great-granddaughter now lives in the Doddy house. Nearby is the Miller Amish Cemetery where Mom's parents are buried and quite a few of her relatives including Jonathan Miller who was the first person buried in that cemetery, (see Jonathan Miller).

Mom had two brothers that died young, Sylvanus died when he was five years old and Ammon died when he was thirteen years old. She talked about them often. When Mom was thirteen years old her mother, Mary died. Later her dad Christian C. married a widow from Holmes Co. Ohio, Barbara Yoder, they had one child that died, Barbara had five children from her first marriage, Abraham, Mary, Levi, John, and Susie Yoder. That must have livened up the household!! Mom talked about the good times they had.

Mom went to Taylor School which was about a mile away, I don't know how many years she attended, I have several books with her name written on the inside cover. Mom, Ada had a twin sister, Katie, and they were identical twins so much so that people had trouble telling them apart, even Dad, when they were courting, one time the girls switched on him and he didn't know the difference until he started talking about getting married, then Katie decided she better tell him who he's talking to.

Young people had a sense of humor and had lots of fun even back in those days, one trick they played on Dad and Mom when they were courting, they took one of the back wheels of the buggy (which was larger) and put it on the front, and the front on back, so they rode all the way home with the buggy at a crazy angle. Another thing the young fellas did, was take one of the wheels off and poke the axle through a fence then put the wheel back on, then Dad had to change it with out getting his good clothes greasy.

Mom and Dad were married on January 2,1908, they probably had a traditional Amish wedding and I assume they were married in her parents house with lots of their friends and relatives attending, with lots of food and fellowship lasting in to the night with group singing Hymns and maybe even some Ballads.

Among Lester's things, we found a pair of socks

JACOB J. C. MILLER (JAKE)   115

with a note attached (written by Nancy) it says these were made by Ada and were worn on her wedding day, January 2, 1908. I guess back in them days girls liked to have some thing a little fancy, at the top and toe, but between dress and shoe what showed was all black. Looking closely you will see they have been darned many times, what with raising ten children and through the depression years they had a lot of use in their lifetime.

They had ten children;

Oba J. b. May 11,1908. d. Jul. 1972 m. Fannie Miller b. Jan. 12, 1911 d. Oct 3,1973 7 children Middlebury, In.

Clarence J. b. Aug 3, 1909 d. Jun. 29,1973 m. Alma Miller b. Dec. 31, 1912, d. Dec. 2, 1992, 5 children Goshen, In

Ada and Jacob

Oba  *Clarence*  *Barbara*  *Mattie*  *Alice*

Olen  *Saloma*  *Nancy*  *Lester & Jo*  *Chris*

Barbara J. b. May 31, 1911 d. Jun. 18, 2004 m. Jacob Hochstedler b. Feb. 6, 1917, d. May 20, 2007, 7 children Howard County In.

Magdalena (Mattie) b. Feb. 15, 1914 d. Aug. 20, 1980 m. Elmer K. Miller b. Oct. 6, 1906, d. Nov. 14, 1984 7 children Shipshewana, In.

Alice J. b. Apr. 12, 1916. d. Sept. 4, 2005 m. Perry Yoder, b. Jan. 12, 1914 d.

Nov. 26, 1941, 4 children. 2nd. m. Jacob Hochstetler b. Apr. 17, 1907, d. Dec. 2, 1979, 3 children. Emma, In. & Sarasota, Fl.

Olen Jacob b. May 30, 1918. d. Sept. 1, 1993 m. Sarah Mae Miller b. May 9, 1917 d. Jun. 5, 2011, 5 children Bristol, In.

Soloma J. b. Jun. 17, 1920 d. Apr. 14, 1960, 2 children. at home

Nancy Alta b. Mar. 5, 1922. d. Sept. 28, 2007 m. Melvin Lambright b. Mar. 2, 1922 d. Jun. 18, 2012, 6 children Middlebury, In.

Lester John b. Nov. 7, 1925. d. Jan. 23, 2012 m. Josephine Graber, b. Jan. 24, 1923 d. Apr. 11, 1965 3 children m. 2nd Mary Schrock b. Feb.11, 1928, Howe, In.

Chris Jay b. Sept. 7, 1930 m. Alice Miller, d. Jul. 3, 2004, 1 child, m. 2nd. Vera (Kauffman) Steiner, b. Mar. 10,1933, (Vera 7 children Bristol & Goshen, In.

I understand when they were first married they lived the first house east of her parents for while, Dad talked about working on a floating steam shovel used to make drainage ditches in that area.

Like most of northern Indiana it was swamp land and had to be drained before it was suitable for farming. Before they had crawler type shovels, the best way to get around was to dig a hole and then float the crane on a barge, then dig it's way up or down stream. This photo shows "Floating dredge on the Little Elkhart River on U.S. 20 east of SR.13"

Today you can see many drainage ditches in that area and I always wonder which ones did Dad work on?

Around 1915 Mom and Dad moved to West Branch Michigan. They moved by train with the family

Michigan train drawn by Velma

JACOB J. C. MILLER (JAKE) 117

going by coach and Dad riding on the freight car with the live stock, Dad talked about how the engineers knowing someone was riding in the cars would give the cars an extra hard bump when they were switching. It was probably a long journey and there were not very many roads in Michigan at that time.

Dad said when they arrived at West Branch, he thought it looked so desolate, that if he had the money he would have shipped right back to Indiana. But they cleared the land and carved a homestead out of the wilderness. And wilderness it was, with virgin forests, wild animals, wagon trails through the woods for roads. Three of Dad's brothers also moved to West Branch, David's, John Jr.'s, and Obbie's, John Sr. also bought land but never moved there

The best cash crop was Navy beans as WWI was in progress and the demand for beans was strong as they were used to feed the military. Dad also worked winter times to cut and load ice into railroad cars, which were used to haul perishables before they had refrigeration units. This was very dangerous work. But ready cash at the time.

They had milk cows, beef cattle, hogs, chickens and a garden so they had plenty to eat but the winters were cold and long with a short growing season, and the soil was sandy with a shallow sub-soil so it was limited to what they could raise.

In 1919 they moved back to Indiana again by train, the children thought that was a long train ride, with Dad again riding with the cattle, farm equipment and furniture.

They first moved to a farm north of Middlebury, The rest of Dad's brothers also moved back to Indiana about the same time.

In 1920 Saloma was born and when she was two years old she became sick and had a very high fever which "burnt out" some of her brain tissue and left her severely mentally handicapped (about the same as a six year old) and was subject seizures all her life, Nancy and Mom took care of her until Mom had a stroke in 1952 and Nancy had married and moved away. When Saloma was about 16 she had twins, Roman and Katie. Katie was sickly and died at 3 months. Roman Jay weighed 4 ½ lbs. and never matured, with a mind of about a 6 month old and had to be hand fed and cared for all his life, he died at 19 yrs old. Mom also took care of him until it was no longer possible. Mom took care of them both all those years plus raising us other children and taking care of regular household duties, and never complained although she weighed only about 98 pounds!!

Mom and Dad lived on several different farms and Dad was always doing some thing with engines and threshers.

Quite a few years my Dad operated a steam engine and thresher for Sidney Zook, I found this photo of Sidney Zook's steam engine

Steam Engine and Threshing Machine, operated by my Dad ??

and threshing machine, I don't know if this is Dad or not but Dad was tall and "stout" and always wore a slouch hat?? Like the guy on the tractor

Dad always had interesting stories to tell about threshing and steam engines, some of which could have been serious, such as when he stepped off the engine and the boiler blew a rivet and it passed right where he had been standing, another time when he was going down a hill towing a threshing machine when the pin jumped out of the hitch by the tractor, so by controlling the throttle he could keep the machine on the road with the tongue between the rear wheels, while keeping ahead so the machine would not pin him against the tractor. They also used to play jokes on one another, one morning as Dad's job was to start the fire and get steam up by the time the guys had their wagons loaded and ready to thresh, this morning he could not get the fire going no matter what he tried, the teams came in with loaded wagons of wheat sheaves and were waiting, while Dad was a nervous wreck and sweating trying to get the fire going, when someone yelled out "Hey Jake why don't you take that bucket out of the smoke stack". The bucket was inside the stack so that Dad could not see

JACOB J. C. MILLER (JAKE)

it from below and smoke could not go up the stack!

Dad was a jolly fellow who could pull jokes and tell stories with the best of them.

In 1934 during the great depression when many people went bankrupt and/or lost their homes and farms, the bank where Dad had a loan and his brother had co-signed, was called in and they had to move off the farm. Olen told me as they were moving out the lane Dad said there stands the wheat, almost in heads already, if I would have been able to stay and harvest that wheat I could have paid my loan. Although it hurt Dad terribly with emotion and regret, he never held any ill feelings against anyone, I remember going to Our Uncle's when I was just real small and I never realized what happened till I was told much later. In old age Dad and his brother were good friends.

They then moved to Lagrange County, near Sand Hill school for a couple of years then moved back on the Rensberger farm on State Road 13, 2 ½ north of Millersburg. The house was actually a log house with siding so that you couldn't tell it was log, they built a large straw shed attached to the barn while we lived there. My Uncle Jake Schlabach came with his steam engine and portable sawmill and cut the boards and timber for the shed from trees that were cut in the woods on the farm. State Road 13 was just brand new and the pavement was white and bright, I thought it was the nicest road anywhere and big? Semi trucks and fast cars used to go past our place.

Olen was courting Sarah Mae at that time and she lived up by Topeka and when Olen came home at

Ole' Rop and Buggy

?? night time he would lay down on the seat of the buggy and go to sleep, and

120   A JOURNEY TO THE FUTURE

faithful horse "Rop" would bring him home, when Rop came to our barn, about ten miles, and stopped suddenly, Olen fell out of the seat and then knew he was home. Rop crossed St. Road 5 and went for a ½ mile on St. Road 13 without a driver!!

Later they moved to the Huffman place, east and south of Millersburg, it was owned by the Federal Land Bank and it was somewhat run down and Dad worked at getting things looking better and restoring the soil to better productivity. Dad was good at using crop rotation plus natural and commercial fertilizer. Stoney Creek ran the length of the farm on the west side and the New York Central railroad ran the full length on the south side. Here is where we had the episode with the stray dog, see (THE MADDOG by Chris), The 20th Century Limited passed by on a daily basis, and the creek bottom supplied us boys with a lot of entertainment, as well as wood for the stoves and pasture for our livestock. When things were fixed up better, the farm was sold and they had to move.

They moved two miles east and north near the Wabash Railroad see (Unlucky Switchman) after about one year they moved south and about 1½ west of Millersburg also along the Wabash Railroad. The railroad actually ran thru the farm and we had a wooden viaduct under the railroad so we could get to the fields and woods on the other side. The barn is still standing, it was never painted and it (stands on rocks) along C.R.44.

Next they moved about three miles east on the same road and in a short lane the farm was owned by Charles Brown and was also run down as nobody had lived there for a while, but when it was fixed up it was a very pleasant place. We could again see the New York Central railroad about ¾ mile away. There was a woods with a sugar camp that we tapped the trees and made maple syrup.

Mom and Dad were Amish all their life just like their ancestors for generations, until about 1950-51 when they left and joined the Burkholder church (now the Amish- Mennonite Church), and bought a car. I'm sure this was no easy choice for them as they had always been faithful to the Amish Church, but now in their old age they had to make that decision. There was disunity in the church, one of the items was about the use of tractors. Dad was always working around tractors and other types of machinery, and he did what he thought was best when the work for the horses was too hard or too hot and stressful and he took pity on them and bought a tractor do the hard work and later another church member also did the same thing. But some from the church did not agree and the differences could not be reconciled. So they left the Amish church.

By that time all their children except Barbara had left the Amish.

We were doing the farming, about 80 acres with a 1927 Model D John Deere that Dad used for threshing and farming. One spring I wanted Dad to get a more modern tractor but he thought this would do, so we spruced it up with a good muffler and I made a nice seat for it and come spring we

Dad's John Deere

found out the muffler wouldn't work and the seat was worse than the old steel seat, and for some reason the engine threw a connecting rod and we needed a tractor to finish plowing so Dad bought a used 1946 Farmall 'B' it proved to be a dandy little tractor and big enough for us. Dad wanted to fix up the John Deere but he couldn't get it running right and couldn't figure what was the matter, it wasn't 'til fall that he discovered that the camshaft was bent, when he fixed that then it ran fine and then he sold it.

Lester was courting Josephine Graber and they got Married in 1947 and they moved to Elkhart, later on Mel Lambright (see Premier single shot by Lester) was courting sister Nancy and they got married and moved away.

About that time I also discovered girls, and I met a nice girl, Alice Miller, at Griner Conservative Mennonite church and we got married on May 4, 1950. We moved into a small house about ½ mile west of my parents by the railroad viaduct and I started working at Larimer's garage in Millersburg. In the meantime Mom suffered a stroke and was partially paralyzed and Dad couldn't handle farming any more so they had an auction and sold their farm equipment and some of their furniture, then moved west of Middlebury, later they moved by Oba's in a converted chicken house which Oba fixed up real nice.

Dad had a small work shop and made things out of scrap lumber for the kids and Grand children. (which quite a few are still around). He also worked at a trailer factory, sweeping floors etc. and hauled some Amish people at times.

In 1959 Dad had a heart attack and died while trying to start an engine to saw wood for a neighbor, he wrapped the rope around the starter pulley and just fell over. Dad always said he wished that when it was his time to go, he wouldn't have to linger long and have people

Doddy's House

take care of him, he was busy right to the last, ready to help someone. He left a legacy of hard work, deep faith, kindness, love for his family and friends, faithfulness to the church with a sense of humor and always ready with an interesting story or a good joke. More importantly his concern for the welfare of his children, raising them in the fear and admonition of the lord. Meaning he wasn't afraid to apply a large hand to my bottom at an early age to bring a point across, I learned that he meant what he said. I don't know how he treated my older siblings, but they always said as the youngest, I got off easy!!

Then Mom went to live with Elmer K's (Mattie) and she also stayed with the other children also until she died in 1963. Leaving a legacy of a devoted wife for sixty-three years, mother and grandmother. Remembered for her hard work and concern for her family, physically and spiritually. Sacrificing her own time for those she loved and those who could not help themselves. I did not realize at the time how hard Mom worked taking care of her family and also the extra work of taking care of two invalids who could not take care of themselves, for many years with out complaint. Until she could no longer because of age and health problems.

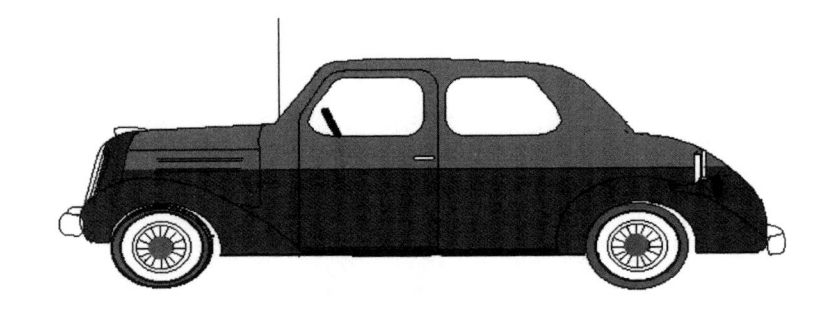

Keep an open mind,
BUT DON'T KEEP IT TOO OPEN OR
people will throw a lot of rubbish into it.

BOOK FIFTEEN

Chris Jay Miller Story

I was born 0n September 7, 1930 a on the "home place" 4½ miles east of Goshen, Indiana. When bro. Clarence saw me the first time as I was laying on the davenport, he remarked "why he looks so cute I think he should be called Christie" and that's how I got my name altho my Grand-father was named Christian and Mom thought we should have a Chris in the family.

Christie 6 yr. old

The Home Place

My first memories are of a living in a big house with my older siblings, none of which were married yet except Oba, I thought it was a happy time until we moved away from there, I didn't know why until later, we moved to Clearspring, Lagrange County, We soon moved back to

Clinton Twp., on State Road 13, 2½ miles north of Millersburg. We lived there several years. State Road 13 was a new concrete highway, smooth and white, the nicest road around. We had a house, barn, drive thru corncrib and other out buildings including an outhouse. They added a huge straw-shed to the barn when we lived there. My uncle Jake Schlabach came with his steam engine and portable saw-mill and sawed the lumber for the straw-shed. I remember Bro. Oba walking the frame ridge peak before the roof was on the framework. Mom couldn't believe he would do that, but he was a carpenter/builder and was used to that.

An incident happened, the first of many throughout my life relating to history in the making and witnessed at the time. On a sunny afternoon when I was about six years old we heard a noise in the sky and it was not an airplane, appearing among the clouds was this huge silver shining object, with pointed nose and tail like an airplane, glittering in the sunlight, It seemed to appear very close but it was far away, Someone knew it was a dirigible, a rigid framed blimp type of airship only much larger, I learned later it was The United States Navy airship U.S.S. Macon which was equipped with a flight deck and could hold a number of small "Sparrowhawk" bi-planes which were lowered and released with a hook device from the ship. The route it was taking at the time was from the east coast to Chicago, then to the west coast, when it reached the Pacific Ocean it ran into a storm and broke apart and crashed into Monterey Bay.

The U.S.S. Macon

Airplanes were rare at that time, whenever one flew over everyone stopped and looked at it. I remember seeing the American Airlines planes, the DC3 had just been introduced, and one day I found a paper cup in our field with the AA logo on it I thought it was amazing, that it had came from one of the planes flying overhead; I guess at that time they just threw things out the window.

A JOURNEY TO THE FUTURE

I started going to school there at Clinton Community School, I had to learn English as well as the ABCs. I had a good friend neighbor boy, Robert Anderson and He came down with polio and died later at about 9 or 10 yrs old, and I really missed him, I named my son Robert.

We then moved to the Huffman place, east and south of Millersburg and I went to Judy school about 4 miles south of Millersburg. My nephew Alvin, who was about my age, and I had many interesting times at the Huffman place, we played with our home-made sand cars (no wheels) and trucks, making roads and scenery in the weed patch. Alvin was always good at making roads they had to be banked on the curves just like real roads. At the back of the farm ran Stoney Creek the full length of the farm and we enjoyed many pleasant times exploring, wading in the creek and occasionally swimming although it was hardly deep enough.

This was starting of the WWII years and I remember hearing Adolf Hitler on the radio one time when my cousin Dan was there with his car, My Dad could hardly believe it and I don't know if it was a direct hook up or delayed broadcast. When Pearl Harbor was bombed and the U.S. went to war with Japan they had a radio at school and we listened to the news as it happened.

On the south side next to our farm ran the New York Central Railroad, at that time all the trains had steam engines and we could see them easy from our house about ¾ mile away. We could see the smoke pouring out of the stack and hear the whistles blowing. The 20th Century Limited was always on time and a daily sight as well as a local mixed freight and passenger train chugging along and stopping at every town. Sometimes when we were close we would wave at the engineer and sometimes at the people in the cars.

20th Century

From there we moved 3 miles east and ¾ mile north of Millersburg right near the Wabash Railroad there was a siding close to our house, see (Unlucky Switchman).we would wave our kerosene lanterns in a vertical circle and that

CHRIS JAY MILLER STORY

would get us a toot from the trainmen. It also had a woods in back and we would make blazed trails thru the woods just like Indians used to.

Next we moved about 2 miles east of Benton also near the Wabash railroad. And I went to Benton School. That is where I first met Alice, Katie. Odena, and Marvin. That barn, (barn on rocks) is still standing although it was never painted as well as the house, the house had an inverted front porch, instead of being attached on front it was recessed into the house. During the war years many things were rationed and hard to get, we could get a limited amount of gasoline and tires. Lester was driving a car by then and I think he used the max. of farm gas that he could get, as he had important places to go. Tires were another thing, he would patch them and patch them again, he said he learned how to remove the tire, patch the tube replace the tire in the middle of the night using only several small screw drivers and a hammer. It could be done if you had to.

The Barn on Rocks

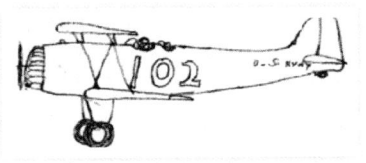

Alvin and I enjoyed looking at pictures of tanks, jeeps and other vehicles of war also battleships and especially air planes, I think we could identify most of the planes, ours and the enemies just by profile. We had to do this in secret as our parents didn't believe in war and I don't think we realized just how bad war was, we were more interested in the machines. Oba's lived next to a practice landing field, not far from Grissom Air Base, built for pilot training, and Alvin could watch them out their back door. He drew some pictures in a letter and sent it to me which I still have. Next we moved about 4 miles east on the same road and again we could see the New York Central Railroad only farther away. This was the Charles Brown place and it was back a lane. It had not been lived in for some time and it was grown up with weeds and brush but when we got it cleaned up it was a pleasant place to live. It had a windmill and pump close to the house and people, ie. The breadman etc. would always get a drink and say it was the best water in Elkhart County. We used to find Indian arrowheads when we were working in the fields.

There was a sugar camp back in the woods and every spring we would gather and make maple syrup, it also had 10 acres of woods and I was in charge of keeping the crow population down, as I had by that time a 22 rifle and I also learned just how smart the crows are, they would sit on a dead tree back by the woods about a quarter mile away and go Caw-Caw at me and I would take a shot at them and he would just go Caw-Caw you missed me!! But sometimes he would jump when I shot so I bet he heard the bullet whistle past his ears! And when it was a Sunday they would come up and fly around the house when we were going to church just as if they knew it was safe. I found out Ole Brier Crow was very intelligent for a bird.

About that time I started to go to Griner Conservative Mennonite Church, I either went with Lester or with the neighbors, the Reigsecker's, I was baptized and joined the church, I also then started driving a car. Dad had bought a car for Lester, but he got into trouble and Dad wouldn't let him drive it anymore. Then Nancy drove it some and then I got to drive it. It was a 1935 Ford V-8 two door that had seen its better days, but with muffler held on with baling wire and blue paint brushed on and silver wheels it was wheels for a teenager. Bald tires and mechanical brakes, which meant the harder you stomped the quicker you stopped but, if they were not adjusted properly you would go to right or left, maybe. But I had fun with it anyway. And no sealed beam headlights either! Just bulb headlights.

'35 Ford

At that time Lester and Josephine (Graber) got married and moved to Elkhart and soon after that Nancy got married to Mel Lambright and moved away. I also started noticing girls and I met a nice girl at church named Alice and later on we got married and moved a ¾ mile west to a small house by the

CHRIS JAY MILLER STORY

Railroad viaduct.

Alice and I got married on May 4, 1950, Alice was born in Iowa and Her parents moved to Indiana in 1936, they later moved to the Millersburg- Benton area and Alice, Katie and Odena went to the Benton school when I was in the 8th grade, that's where I first met Alice not knowing I would spend most of my life with her.

On Alice's first birthday as she was learning to walk, she walked towards a kerosene space heater and put both of her hands on the very hot heater, she burned both of her hands very badly and she had those scars all her life.

Alice had Rheumatic fever when she was young, (she found out much later that she was born a "Blue Baby", she was born at home and her Dad was the mid-wife and he told his sister that he got the cord unwrapped from her neck before she turned blue, but I guess not quick enough). When Robert was born the Doctor told us she should not have anymore children and her life expectancy was about 45 yrs., but she made it to 73 yrs, and was able to see and hold her first Great-Grand Daughter.

Alice's parents were Clemen and Mattie Miller, they had lived in Kansas at one time then moved to Iowa and later to Indiana. Mattie was related back to the Christian (Christal) J. Miller's son John C. Miller who had moved to Haven, Kansas.

Clemen C.B. and Mattie had fifteen children, all born at home and all but one lived to maturity.

Lizzie C. b. 1910 d. 1942 m. Menno E. Bontrager Elkhart, In.
Joni M. b. 1912 d. 1970 m. Mattie E. Bontrager Macon, Miss.

Barbara b. 1913 d. 1996 m. Joe E. Miller Middlebury, In.

Fannie C. b. 1915 d. 2009 Millersburg, In.

Mahala C. b.1916 d. 1996 m. Henry J. Yoder Kalona, Ia. and Sarasota, Fl.

John C. b. 1918 d. 1971 m. Emma Jane Corwin Yakima, Wa.

Mary Ann b. 1920 d. 1990 m. Abner D. Miller Plain City, Oh.

Nettie C. b. 1922 d. 2011 m. Samuel Ring Elkhart, In.

Eddie C. b. 1923 d. 2008 Mishawaka, In.

Freddie C. 1923 d. 1925 Kansas

Mattie C. b. 1925 d. 1985 m. Harvey E. Miller Bristol, Ind.

Clemen Jr. b. 1929 d. 2002 m. Polly Eash, m. 2nd to Mary D. weaver

Alice C. b. 1931 d. 2004 m. *Chris J. Miller Bristol, In.

Katie C. b. 1932 m. Lloyd Dalton Troyer Mio, Mi.

Odena C. b. 1934 d. 2014 Goshen, In.

After Alice and I got married, I started working at Larimer's Garage and Shell filling station. In Millersburg. Which started my career as an automobile mechanic, basic mechanics came easy, I guess it was in my genes, from my ancestor Chrisian "Schmidt" Miller on down. But when I wanted to advance to automatic transmissions, that was a different story. So I took a night course from Lincoln Technical to service automatic transmissions which was quite new, as only about half of cars had them. Also there were not that many guys that could work on them. My first transmission specialist job was at Culver's Oldsmobile and Cadillac dealership, Elkhart, I worked at many different auto shops, including Nash and Studebaker! last I worked at Max Myers Chevrolet in Middlebury, as an all around mechanic, doing every thing from front end alignment, major overhaul and automatic transmissions. But with dirty busted knuckles and a box full of expensive tools and poor pay I decided to give it up.

My brother-in-law Harvey Miller was working at Adams and Westlake in Elkhart and he said they were looking for a maintenance man to work on steam, air, water, and gas, and it paid better than what was getting. It was a Union shop I didn't care for that but I worked there about six years, I gained several merit raises and was a special service maintenance man in their 'white room', a special ultra clean room where they made mercury whetted relay switches that were used by railroads and early computers. I enjoyed the work, I had to wear a white gown and safety glasses while in that room. I helped to build and maintain machines that used high pressure Hydrogen and Nitrogen gasses which was interesting, and interesting people to work with, I couldn't understand the working of the union shop and when they said we had to pay much higher union dues because the auto industry in Detroit was on strike and they were running out of money (but Jimmie Hoffa was not changing his life style). So I refused to pay my union dues and I got terminated but with good standing in the company, and they were trying to keep me working there to no avail.

In 1962 Alice and I bought a house on Barbarah Drive south-west of Bristol, In. where we lived for 45 years. We enjoyed the quite wooded area with dogwood trees blooming the spring and colorful leaves each autumn, then we had to rake them up and dispose of them!!!, Alice always enjoyed gardening as long as she could and tending flowers around the house. We had pleasant neighbors who would watch our house when we were gone and many good and enjoyable memories of our home.

On April 11, 1965 a whole series of tornados raked across Indiana killed and injured many people and doing millions of dollars worth of damage, at least 4

or 5 funnels went through Elkhart county. My Brother-in-law Harvey and Mattie Miller's house was destroyed near St. Rd. 15 and U.S. 20. They lost nearly everything, but luckily nobody was at home. Another one struck north of Middlebury, where Lester and Josephine had a new house about a year and a half old, Josephine's parent's Simon and Mary Graber also had a new house across the road, and across the river from them, Lester's and Joel Graber's were there for Sunday dinner, when one tornado went through just north of them, Lester, Eugene and Joel went to see what damage was done and when there they looked around and saw another tornado destroying both houses, Lester's and Simon's where the rest of the family were. All of them were injured and Josephine was killed by beam falling on her head. Both houses were completely destroyed, cars wrecked and Lester's pick-up (that Him and Me had just tuned up the night before) landed in the river. Lester was able to salvage a lot of his things. A lot of people in the area had major damage done also. All communication was down, they shut down the electric grid to keep people getting shocked from all the power lines down. The weather bureau issued a all points tornado warning because they could not keep track of all the tornados.

Double Tornado between Goshen & Elkhart

A night to remember!! I had been called to Adlake where I worked because they had a roof leak, I went in and on the way home, on the radio they were asking anyone with a station wagon to come to the Dunlap area to transport injured people, so I thought this must be bad, and when I got home Alice had left a note saying she was with Harvey's and I should come over. When I got Harvey's what a shock, every thing was destroyed, their house was leveled, but no signs of any cars, so I waited awhile 'til they came back and when they did they had heard that Lester's house was hit and they thought that Josephine was killed, so we decided to go find out. Many roads were blocked from trees and debris, the main roads were closed to keep unnecessary traffic off. So by taking back roads and many detours we came to Lester's place, they told us all of them had been taken to Goshen hospital. Then we wound our way to Goshen, at the hospital it was a place of mass confusion, they had a desk in the entry way where they would try to keep record of every body, they told us Josephine was dead but Emily was in the hospital as well as Simon and Mary Graber and some of the Grand children, we asked to see Emily but at the desk they said they don't have a record of her, but they said why don't you go see if you can find her, with a little search we found her under a different name, and we also found Simon and Mary Graber, they didn't think that Mary would last through the night, but she did and lived quite a few years.

Harvey and Mattie's house

Then it was the funeral to go to, the grieving and letting go. Then the reality of cleaning up. Restoring and rebuilding. Lester and Simon's houses, being just new and insured would be rebuilt on the same foundation. But Lester needed emotional support because of the sudden loss of his spouse. And I tried to be with him as much as I could.

But Harvey's had to get a loan (which was no problem) for materials and then rely on volunteer help, which mostly built their new house. The volunteer's that showed up to help was unbelievable, a God send, we would not have been able to recover as quickly as we did if it hadn't been for them, a group came from Pennsylvania with their equipment and built the kitchen cabinets for Harvey's house, We moved our mobile home over to Harvey's yard so they would have a place to stay while they built their house. Many emotions to go through at a time like this, there was a before and after wards, your life and many lives were changed on that day.

    Later on after I had left Adlake in about 1968, I met a friend, Joel Graber and when he found out I wasn't doing anything at the time, he said why don't we start a trailer factory? And I said why not. I didn't know anything about building trailers or starting a company/corporation, but we had some money saved up and I was adventurous, and we began. We got a third partner, Marvin Summers and formed a corporation, we called it NORTHWIND Manufacturing, Inc. I thought they knew how to do things and it turned out they knew less than I did. But we got set up and started making travel trailers and truck campers, things went fine for about a year and we had orders for units but got in a slump where we didn't have cash to buy materials, so we decided to take on another partner who promised he would supply cash and let us run the business, but it didn't work that way as he had a successful business but he didn't understand the trailer business, which is unique as far as business is concerned and must be run accordingly. Joel decided he didn't want anything to do with it anymore and took a pay off and left.

    I decided to keep my stock but would not work for the company, it wasn't long until they were in financial trouble again. And they came to me and asked if I didn't have some relatives that had money I could borrow from and I said yes, but I'm not going to, I could see the company was not going to survive and I was not about to send good money after bad especially if it wasn't mine, soon the company closed up for good and some money lost and lessons learned. And I moved on.

    I got a job as a purchasing agent at Saturn Campers in Wakarusa, I worked there for several years and enjoyed it very much and made good money, they gave me free rein as to what to buy but I had to show a profit at the end of the year.

    Alice and I were thinking of buying a camper ourselves but couldn't decide what to buy. Travel Equipment in Elkhart was converting vans at the time, they would cut the tops out of vans and made a lifter system that raised the roof about 20 inches so you could walk around inside real good, they installed a camper kit for the van, but I didn't like their floor plan and besides I wanted to do it myself. The van we bought was a 1969 Chevrolet, this was the last year where the engine was between the seats and you actually sat over the front wheels, it looked like a small bus in front, it had a V-8 engine, automatic transmission, but no power brakes or power steering.

Alice was intimidated by it at first, she thought it would drive like a truck, but once she got onto it she liked it better than a car. I put vinyl on the inside, put in a heater, refrigerator, stove top, a sink with simple water system, I made a flip-flop seat with hardware that had two seats and a table with a United States map on it, and then it laid down for a bed which was quite comfortable, many times we were asked to see the inside and we got many compliments from it, as for that time it was as nice as any on the road. And it was comfortable for two people.

After I got it done, I told Alice why don't we take time and just go traveling with it, Robert was at home and he could take care of the house and we could explore the country and visit relatives as we want, she was not too keen on the idea as she was not too sure how we would survive, I convinced her I could work and I was not too particular what kind of work I do, so we both quit our jobs and loaded up the camper, (I found out later she had stashed a box of home canned beef, at least we would have something to eat!).

We visited relatives in Iowa then wandered north to Minnesota took Rte. U.S. 2 west 'til we got to Washington state, and we stopped at Alice's brother Johnny's for

About a week, they lived in Yakima, which is the beautiful Yakima Valley, noted for apples of all sorts, fruit trees and hops that are grown to make beer. It is also semi-arid desert on one side and Rocky mountains on the other side, we went camping on Mount Rainier with Johnny's and their daughter Shirley and family from Seattle.

Then we headed west again to Seattle then we followed the coastline south to Oregon where we looked up Alice's Niece Vera and Harvey Bontrager, and we stayed there a bout a week. They lived up Raccoon Valley, across a rickity bridge with a horse barn and small house tucked right next to a small mountain, it looked cozy to me.

When we were there we received word that Alice's brother Joni Miller, in Mississippi was killed in a truck/tractor accident, we knew we could hardly make for the funeral, and as we had planned on stopping there later I told Alice we would visit with them as long as she likes when we get there.

From there we followed the coast south to Eureka, California, then we headed eastward across the Rockies and into the desert across to Winnemucca, Nevada.

When we traveled we done a lot of sight seeing, stopping and roaming around as we felt like it, if we saw a mountain or something in the distance we would take off across country until we got there. We explored many back roads, gold mines, ghost towns, and shops of all kinds, Alice discovered there were grocery stores along the way and we stopped at corner fruit markets and a fish market in Seattle where we got a salmon steak that was just delicious by it's self. Alice found that by stopping at a store every several days we always had a good selection of fresh food and a variety of local food and it didn't cost that much, with gas at avg. 35 cents per gallon and camping was either free at local parks or 4.00 to $5.00 at campgrounds with showers and laundry, so we lived pretty good and clean!

From Winnemucca we headed south to Lufkin, Texas where Alice's brother Clemen lived at the time. We also stayed there about a week and went sight seeing around the area.

Then we headed east from there towards Macon, Mississippi, where Alice's brother Joni and his family had lived for some years Joni was a minister in the Iowa Conservative Mennonite Church and they moved down there to establish a church in the back country rural area. Several other families some from our local Indiana area that we knew also moved down there, to help establish a church community.

Joni was killed when a gravel truck came from behind and ran over him. Joni had a Ford 9n tractor with a brush hog on the back and truck ran right over him and mangled him so bad they could not have a viewing.

We stayed there a couple of days and then we headed towards home we stopped in Nashville and took in a Grand Ole Opry show at the old Ryman Theatre. When we got out it was very late and we had to drive a ways until we could find a place to park overnight. The next day we drove home tired and flushed, but we had lots of photos and lots of memories.

The reason I wanted to take Alice on a long and extended trip was because the doctors had told her that she probably wouldn't make it past forty and I was afraid she wouldn't be around for retirement, so I was glad we could do this when we did. And it was very enjoyable to be just the two of us out in the world by ourselves, sometimes it was scary, sometimes funny, but mostly enjoyable and I don't regret a moment of it. As it turned out it would never have suited to make such a trip later on.

Some time after we came home I got a job at Jayco Travel Trailers, Lloyd Bontrager the owner was childhood friend of mine and Bertha his wife was a friend of Alice's we went to the same church and lived in the same neighborhood when younger. Lloyd put me to work driving a truck delivering trailers well I always wanted to drive truck, but his trucks at that time were a sorry lot of used trucks and equipment but he keep replacing with new and better equipment and also the interstate roads were not completed which sometimes made things interesting, an interesting things happened, I was delivering a load of campers to a dealer in the Miami Fla. area and as I was going down the Florida turnpike, I heard on the radio, they were going to launch a space rocket so I stopped at a rest area and waited and listened and they had another delay, so I decided to keep on driving. The truck did not have a working radio so I carried with me a portable radio. As I was driving along they said they were going to launch the rocket at any time so I pulled over on the berm and walked to the back of the truck with the radio on my shoulder, soon I heard them start the count down and a blast-off which I could hear on the radio, several seconds later I could see the rocket rising from the horizon and watched as it gracefully arced across the sky and about my 2 o'clock position they released the booster rocket in a shower of sparks and flame as it fell toward earth and the rocket continued on until out of site .

This happened to be the Apollo 17, which was the only nighttime Apollo launch, and was also the last moon landing they made, which I witnessed just by chance.

I also worked as a dispatcher, and was a foreman for a while then went back to driving truck again at that time they had diesel trucks and the interstate system was mostly done. Jayco had a manufacturing facility in Harper Kansas and one in Listowel Ont. Canada, they leased a new International Transtar cab-over tractor to transport parts to the facilities, now that made truck driving fun! I put on over 300,000 miles on that truck. Driving truck was the most exciting job I had even tho it got boring at times it hardly ever dull.

In 1978 I decided to quit Jayco and later I got supposedly a part time job to drive a truck for a friend, who wanted to farm during the summer and drive in the winter time. Only it didn't work that way, he couldn't get approved to drive for Sawyer Transport because he didn't have experience. So I got stuck driving full time, this involved transporting dry goods from point to point nation wide. So that meant that some times I didn't get home for a couple weeks at a time.

And that soon got old, but I drove for him several years then resigned.

There was an empty building at the corner of U.S.20 and St.Rd.15 I thought I would like to open to open a lawnmower shop, *Green Gables,* so I leased it and started in repairing mowers at first but that just didn't seem to work, hard work and poor pay. So I started putting in more new merchandise and parts, like Ariens and Yardman lawn equipment, Greenmachine chainsaws and weed trimmers. Later on I started handling Manco go carts (all terrain funcarts) that made selling interesting and profitable.

In the meantime Robert and Rachel had moved to California and Alice started having health problems, in 1988 she had a mitro valve repaired at Cleveland Clinic.

I tried to sell the business but I discovered I had nothing to sell but hard work and nobody wanted that, so I called my nephew, Leonard Miller the auctioneer and made sale and sold everything and retired, I thought, but Manco called and wanted to know if I wanted to be their sales rep. for southwest Michigan and part of Indiana, I done quite well with them for about four years when they were bought out by another company and they decided to sell through mass merchants so they didn't need me any more, unfortunately that didn't work, the go cart business didn't last.

I had to have heart by-pass surgery in 1993 and Alice was having more health problems, in and out of the hospital and the last two years had to have constant care at home and she died on July 3, 2004. after a 54 year marriage. A loving wife and mother with a ready smile and a helping hand where needed. Working at Brenneman Memorial Church children's nursery for over twenty years, during church taking care of infants so mothers could attend church.

Alice and I have one son Robert (Carol), one Granddaughter, Rachel (Justin Minick) two Great-granddaughters, Roxanne and Charlotte Minick. They all live in California now.

Several years later I met a nice lady from Ohio, Vera Steiner and we got married on April 7, 2007, that May I had an auction and sold most of my things and my house and moved to RedBud Court in Greencroft, Goshen, In. with Vera. Vera has seven children, Terry Steiner, Roger Steiner, Rhoda Mininger, Debra Fetzer, Phil Steiner, Sam Steiner and Eugene Steiner.

Rhoda and Eugene live here in Indiana and the rest live in Ohio, Holmes county area, nineteen grand children and ten + Great-grand children.

In Dec. 2007 we took our van with a stereo cabinet (that Robert and I made) plus other things for Robert, and spent Christmas and New Years with Robert and Carol, from there we went to Phoenix, Az. and stayed with Vera's sister Palmer and Joan Steiner, from there to Mission Tx. and stayed with Vera's brother-in-law Roy and Vesta Steiner then to Macon, Mississippi and stayed at Roy and Tillie Miller's and Roland and Sandra Miller's (two of Alice's nephews) from there to Tallahassee Fl. and stayed at Monroe and Ruby Hochstetler's (Alice's niece) then to Sarasota, Fl. Where we stayed 'til April then went home.

We went through 16 different states and visited 14 different churches, we only missed one Sunday going to church because of illness. Now we spend our winters in Vera's mobile home in Tri-Par Estates, Sarasota, Florida

I thank God for a full and varied life, with ups and downs. I made lots of friends over the years that I am thankful for, I am thankful for the memories as I reminisce about the years gone past. I am thankful for lots of relatives, nieces and nephews, all the in-laws. I have said goodbye to Alice and all of my siblings and lots of other relatives and friends that I miss dearly. But I have lots of memories which I am thankful for.

Uncle Chris

CHRIS JAY MILLER STORY

A HAPPY JOYFUL SPIRIT SPREADS
joy everywhere
A fretful spirit is a trouble to ourselves

and to all around us.

A JOURNEY TO THE FUTURE

BOOK SIXTEEN

Robert Dean Miller
by Robert Miller, July 2016

I was born on September 12, 1950 to Chris and Alice Miller. I graduated from Jefferson High School in 1969, then went to Purdue University for one year. I worked for Northwind Campers (my dad's company) doing drafting, and designing. I worked at Starcraft as a draftsman and designer for RV's and boats. While employed there I designed a conversion van that was quite successful for them and also designed the décor for their boats. I also worked for Midas Campers and Coachmen Industries.

On Oct. 31, 1975, I married Carol Widner (Wildman), Carol had a daughter Nikki, and Rachel Alea was born May 17, 1977. Carol and I divorced in the early '80s. Rachel and I first moved to California in 1984; a year later we moved back to Indiana for a couple of years. When we moved back to California I took Rachel & her sister, Nikki, in my little Dodge Colt pulling a huge home-made utility trailer with all of our stuff. While driving across the desert (in

August without air conditioning like the pioneers of old!) two of the trailer rims split because of all of the weight.

After arriving I started working for Omnica Product Design & Development in Irvine, where I still work, nearly 30 years later, where I get to design many types of products, mostly medical instrumentation.

I met Carol Roberts in 1991 and we got married in 1993. Carol has two children Juli and Erik. Carol and I live in Costa Mesa, CA.

Rachel married Justin Minick on May 1,1999 and Roxanne Eva was born Jan. 20, 2000. Charlotte Rose was born Feb.25,2009.

My mother (Alice) died on July 3, 2004. We try to go back to Indiana once a year, and Dad comes out about every other year to spend Christmas with us and Rachel's family. We like to go on cruises when we're not visiting family. In 2015 we finally fulfilled our dream of going to Europe by taking a river cruise up the Rhine River from Amsterdam though Germany to Switzerland. We also spent a few days each in London and Paris.

Robert and Carol

Carol retired from the University of California, Irvine, and I'm thinking of (semi) retiring in a few years. We plan to stay in Costa Mesa in our house that we've remodeled to our liking, including adding a 'doddy house' master suite that I designed for Carol and me. We've been "environmentally aware" and pro-active for a number of years. We got 32 solar panels in '03 and may be supplementing them soon. They were followed shortly by a complete change to all Orange County, CA native landscaping (requires no water to speak of). We've had 2 all electric cars and one hybrid car. We plan to get a higher-range EV in 2017!

(Robert wanted me to write his story)

In the summer Robert was sixteen, he wanted a summer job, so we got him work at his Uncle Lloyd and Katie (Alice's sister) Troyer's they had a dairy farm at the time near Mio, Michigan. This gave Robert a brand new experience in life, (a city boy going to a farm) but he did learn how to drive a tractor, a pickup truck and how to milk cows and take care of other farm animals, which was a good learning experience! He also had a bout with appendicitis, which was scary as when Lloyd's took him to the hospital they wouldn't operate until they contacted his parents, they couldn't locate us until the emergency passed, so he didn't have the operation.

When the summer was over we told him that we would pick him up and go up to Montreal, Quebec, as the World EXPO 1967 was going on. We had our boat on top of our '64 Chevy station wagon along with a tent and camping gear, we were planning on camping and fishing along the way, which we did in Ontario.

When we got to Montreal we got a campsite on the north side of town, Just as we had our tent set up and Alice and Robert were stashing our gear in the tent, up came a rain storm, a real gully washer, the water was flowing under the tent and soaking our sleeping bags and some of our clothes, Not good.

As planned, we would take the "Metro" (subway) to he EXPO, seemed like a good idea. Quebec being mostly French speaking, they had signs in English pointing " To the Fair" but coming back every thing was in French, whoa! Robert realized this as we left our station and by calculating from the next station we marked which was ou station to look for when we come back.

We enjoyed a full day at the EXPO especially Robert he was fascinated by all the new technology on display from all parts of the world. Alice and I enjoyed all the buildings and food from other countries. When we went on the Metro we found our station alright but we were underground with many different stairs leading to different exits. We were being ignored by people (rush hour) when finally someone realized we had a problem and stopped to help us even thou we could not understand each other, we showed him our parking ticket

ROBERT DEAN MILLER

and he showed us to our car.

From there we went down through New York State. Where we camped one night at " The hidden Campground" right beside what we thought was a hill which was really a railroad embankment and during the night a train came along and we were sure it was coming right through the tent. Scary!

We traveled the rest of the way home with incident, but many memories of an enjoyable trip. We had many other interesting camping trips Alice, Robert and I, in Michigan, Canada and Indiana

And one trip we went to South Dakota badlands, Mount Rushmore and the Crazy Horse Monument when they were just starting to carve the mountain.

With station wagon and tent and later with our van camper.

I always enjoyed camping or fishing with my family, or just exploring and traveling, To reminisce brings back many happy memories.

Chris

BOOK SEVENTEEN

Rachel Miller - Minick

I was born in May of 1977 to Robert and Carol (Wildman) Miller. At the age of 5, Mom and Dad divorced. My Dad and I moved to sunny California after the divorce was complete.

I was blessed with frequent visits during summer and holidays to see grandpa Chris and grandma Alice via airplane. It was an exciting adventure to fly and especially fly by myself. Robert would put me on the plane in Los Angeles with a tag tied to my dress that had the name and address of who was supposed to pick me up.

Then Grandpa's would pick me up at O'Hara Airport in Chicago. Then they would send me back the same way. However when I flew back home to California, I would usually arrive home in California before grandpa's returned home from O'Hara Airport to Bristol, Indiana.

I went to culinary school in Vermont where I met Justin Minick. We married soon afterwards in Seattle, Washington. After which we moved to Vashon

Island where Roxanne was born in January of 2000. (Home birthed on the island with no road off, and grandpa was worried!).

Later we moved to Woodstock, Georgia, close to Justin's parents. And after many hot blistering years in Georgia, we cooled off in Vermont when Justin got a job with the State of Vermont servicing air conditioners. This is where Charlotte (2nd child) was born in 2009. Soon after Charlotte was born Justin's job was eliminated with the State of Vermont due to cutbacks. We then moved to Laguna Niguel, Calif., where Justin also serviced air conditioners.

From there we moved to Felton, California, where we now live with a half –mile long driveway up to our house, over looking a large valley. Justin has his own business now installing and servicing commercial heating and air conditioners. I work at raising two daughters and I also own my own business. Roxanne works part time as a wood worker and wood shop helper and Charlotte attends a local grade school.

Roxanne, Rachel, Justin and Charlotte

We like gardening, hanging out with our chickens, traveling with the girls, new adventures and visiting with grandpa, Vera and family when they are brave enough to visit our valley.

I found inspiration and a brief time to write for your book and this is what inspired me;

I loved the yearly rituals and ceremony that grandpa Chris and grandma Alice introduced to my life. The greatness of ritual supersedes everyday rhythms. From the window of today the viewing spectacles of years past create the essence of my life and the timely traditions passed down for generations to come.

My grandparents open hearts and open hands welcomed all who needed love and an extra bit of cheer. My grandma Alice taught the value of quality and the simplicity of midnight talks over the dinner table and a hot cup of tea. Her generous ear taught me how to listen to my own children without judgment and how to love unconditionally. Where as my grandpa's fun sense of humor and adventure for life and learning taught me an invaluable sense of how to live life with a thirst for experiencing everyday to it's fullest.

My memories of summer sandwiches filled with grandma's homemade bread and butter pickles and freshly picked garden tomatoes changed my hatred for vegetables into a lifelong love of gardening and fine veggies. I spent countless hours with my grandparents, my teachers of love, patience, sewing, crafts, puzzles, Sunday dinners after church, hours of monopoly, hospitality, religion and peace.

And so it is, my grandpa has asked me to write a few words recapping my favorite memories of my youth. And upon first observation it occurred to me that maybe my life wasn't positively charged enough to share in detail my life's happenings. Upon further reflection, I have discovered that my life offers a well of fantastic adventures in which to share. As to whether any are of interest to the readers is completely out of my hands and will absolutely involve divine intervention.

So the memory I have chosen are a scattered bit as my memory is not clear or set on just one, but instead what comes to mind is the many summers and vacations spent with my grandparents at their lawn and garden shop, Green Gables.

My dad and mom's divorce brought forth my dad and I's move to California. From this point on I was transported back and forth from California to Indiana spending summer and winter vacations with my grandparents. At the time my grandparents owned and fully operated Green Gables. And what a happening place it was. I loved going to the shop in the early mornings and opening up the doors to the go karts, lawn mowers, 3 wheelers and chainsaws. There is an unprecedented joy opening up something in the morning and awaiting its full opportunity to spring forth each day, and this is how I saw the adventures at Green Gables.

There were 5 major sections of Green Gables. The office, this is where I spent hours with grandma watching TV, helping, building things, using all of their printing paper and more. The faux wood paneled 10x10 room was as glorious as could be for a 8 year old with a vivid imagination.

Section 2, the counter. Also a magical space. As this was where all of the chit chat happened. All the customers and visitors came to the counter to check out, ask questions and…gossip. Bar stools sat on both sides of the counter, just in case someone wanted to stay for a while and chat, and yes they did. On the other side of that sat a full smorgasbord of candy bars and sodas. I soon understood the meaning of powered by sugar. The superhero effects that Mountain Dew provided me helped me move mountains…and lawn mowers.

Section 3, the back. This is where the repairs occurred. Lawn mowers, go carts and more filled some 1000 Sq feet of space with a symphony of roaring engines, the sweet smell of shop grease and a variety of machines. From the counter, a ramp led you into the back of the shop and once in the back, a small pathway led you straight on through to the large garage door that let out the completed work and welcomed in new projects.

Section 4, the exterior of the building. I currently live in California. After many years of bouncing around the United States I have found that I am sucker for good weather. And most of California offers mild winters and an occasional mild summer. Around the San Francisco area, land is not as plentiful and is definitely not as inexpensive as in Indiana. So looking back at the amount of exterior space that existed around the Green Gables it seems it was plentiful for a business on the corner of In.15 and U.S.20.

The front half of the building's outside space consisted of fine pea gravel that coated the entire ground, at least all that was not filled with grass. The grass sections were a necessity for a lawn and garden shop. Why you may ask?

Well, what is the first thing that comes to mind when you see someone mowing the lawn while sitting in traffic waiting for a stop light? Maybe wow, that's a nice mower. And wow that riding lawn mower is so easy to operate a little girl can do it. Yup. That was me. Grandpa's helper, and test driver.

After some convincing that I wouldn't hurt myself grandma Alice gave in to our pleas to allow for me to ride go karts, 3 wheelers and lawn mowers. My school months consisted of the city life. In which I lived in Orange County, CA. with my father in a one bedroom apartment. It was loud and busy and I was happy to escape to Indiana's green grass and the freedom of backyard go kart jumps, speeding around the store grounds and mowing the lawn.

Section 5, the new equipment. Most of the shop consisted of brand new go karts and lawn mowers that were delivered or grandpa and I would go pick up in Fort Wayne In. There was a rhythm to the showroom. As it was a rather small space and shaped like a L. With the back portion leading way to a garage door that allowed for everything unable to fit through the front door to be driven or pushed in and out. The synchronicity came from maneuvering the machines in and out as to not scratch or dent any of the goods. The shiny colored fiberglass that surrounded the engine's exterior turned the cement gloom of the shop's interior into a showroom of color and brilliant designs.

Most of you that recall Green Gables may recall the perfectly aligned go karts and lawn mowers that decorated the lawns. Upon good weather days the go carts and lawn mowers came out of the garage in the early morning and went back into the garage in the evenings. Upon days with drastic weather, sometimes they would never go out. But this didn't stop myself and others from sitting upon and pretending to drive all of the wonderful types of machines.

RACHEL MILLER-MINICK

I looked forward to the time I shared with my grandparents during my vacations. Years later when my grandma Alice's health declined and Walmart moved to town they closed up shop. And even though that once green building is now painted a sandy brown I will always see it as Green Gables, and it will always be the most magical lawn and garden shop ever.

Rachel

If you pray,
YOU DON'T HAVE TO WORRY,
if you don't pray, it does you no good to worry.

BOOK EIGHTEEN

Many Miller Memories

Stories and memories of our parents and ancestors as we remembered being told, some when we were young and not sure of all the details and facts, but stories never the less, stories from long ago, shared by Miller families, friends and other story tellers.

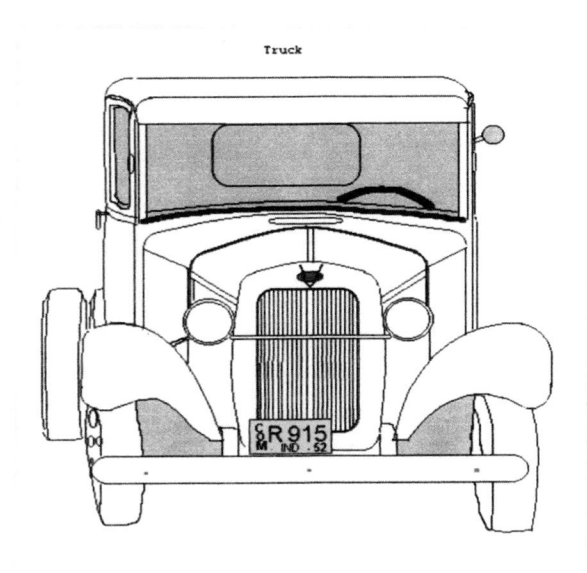

Truck

Old friends are like old trucks, you like to keep them around for the memories.

chris

WHEN WALY-MART COMES TO TOWN

By Chris, about 1985

Bargains, bargains galore

Throughout the store

If you want it or you need

Take no heed.

Gifts and gadgets, small or large

Will this be cash or charge ?

Open all night and Sundays too!

Every thing on sale, bargains you can't resist

But the things you need just don't exist !

Sorry we're out with a shrug

Clerks are friendly but…

Faucet washers in a pack

None to fit mine, from way back

Big Blue- Waly and K sell for less

I know why, Can you guess ?

Charlie, Joe, Ed and Pete

They're all closed down, They can't compete

Grocery stores – filling stations

Run by Mom and Pop

Now You never see

In there place it's a "one stop"

They were our friends and neighbors

They with families, kids in school

With house and car and dreams too !

They paid taxes like you and me,

On the school board, maybe trustee too !

We thought their prices were high

But they checked the tires

Carried groceries and said "Hi"

We thought they were getting too rich
Now our money goes to the filthy rich !
Where does it go, does anyone know ??
New York, Chicago, China, Tokyo ??

A TRIBUTE TO MY PARENTS (NANCY)

Jacob J.C. and Ada C. Miller. Dad was born November 11,1885, and Mom was born in July 19, 1886. They were married on January 2, 1908. They had a large family of 5 sons and 5 daughters. We were all loved and cared for, and our daily needs were met as well as our spiritual welfare. They gave us all spiritual guidance, and we were taught to always be faithful to our living God, and we were free to attend the church of our choice.

Mom and Dad were raised in the Old Order Amish Church both were baptized and they were married in the church. Dad's father was an Old Order minister, and Mom's father was an Old Order Amish deacon. In their later years, they joined the Beachy Amish church.

They survived 2 world wars, the great depression, and they always said they wished they had more material things for us, but we never went hungry or lacked the things we really needed. What we did have lots of was love and happiness. We had fun playing together and were taught to pray and give thanks for what we did have.

They had moved to West Branch, Michigan from March 1915 t0 1919, which they said was not the best move. In the depression years, they lost their farm east of Goshen along with 3 of our neighbors this was a very sad time. It was the first time I heard Dad cry, which I will never forget.

My sister Saloma was born a normal baby, but at the age of 2 she became very ill with a high fever which "burnt-out" some brain tissue and left her like a baby again. She did learn to walk and talk again in time, but she was always a "special" child who we took loving care of. She was "raped" (at about sixteen), and her twin babies were born premature. The little girl Katie Mae, was sickly, weighed only 3 pounds and died after 3 months. The boy Roman Jay weighed 4½ pounds and never matured. He also was a "special" child and died at the age of 19. we all helped in different ways as best as we could, but shamefully the Old Order Amish church never supported them very much, which was hard to take at times. The folks felt they were often looked down on, but they said that if this was "The cross that they must bear", then they would be patient and bear it.

Mom had a stroke in August 1952 which really changed things. Roman had to be taken to Fort Wayne to live in a children's home, and Saloma moved to an Epilepsy Center. They both died in these homes. In his later years, Dad had heart trouble from being gassed while fixing a corn shredder. By this time they had a car, however so it was much easier for them to get around and go places. In spite of their troubles, they remained faithful to their Lord and their church until their deaths. They were always concerned for the spiritual well being of their children, grandchildren, and great-grandchildren.

The topic for Dad's funeral was his belief in God and concern for the welfare of his family, The topic for Mom's funeral was her strong belief and "she did what she could" Mark 14:8a.

Amazing Grace, Will the Circle Be Unbroken, and Precious Memories were the songs sang at both funerals. These songs were picked by Oba and Clarence.

Nancy Lambright

AN OBITUARY (LESTER)

Mrs. Ada C. Miller died suddenly of a stroke at the home of Elmer K. Millers on Feb. 18, 1963 at 12:15 A.M. Was born in Lagrange C. Ind. On July 19, 1886 Dau. Of Christian C. (Mary Bender) Miller Age 76 yrs. 6 mo. 30 days.

On Jan. 2, 1908 She was united in marriage to Jacob J.C. Miller, who preceded Her in death on Nov. 7, 1959 Also a daughter Salome on Apr. 14, 1960 and 7 grandchildren and 1 great-grandchild.

She leaves to mourn 5 sons and 4 daughters. Oba A. Miller, Middlebury, In., Clarence J. Miller Goshen, In., Mrs. Jacob J. (Barbara) Hochstedler Kokomo, In., Mrs. Elmer K. (Mattie) Miller White Pigeon, Mich., Mrs. Jacob E. (Alice) Hochstetler Ligonier, In., Olen J. Miller Bristol, In., Mrs Melvin J. Lambright Lagrange, In., Lester J. Miller Middlebury, In., Chris J. Miller Bristol, In.,

44 Grandchildren, 30 Great-grandchildren, 2 step-brothers Levi Yoder Mt. Eaton, Oh., John Yoder Wyoming, Del., 1 Step-sister Mrs Albert J. (Mary) Wingard Fredricksburg, Oh.,

Her twin sister Katie (Mrs Enos Bontrager) proceeded her in death 44 yrs. 2 days, Also Mrs Dan B. (Mattie) Miller, Mrs Oba J. (Ella) Miller, Mrs Jacob S. (Lydiann) Schlabach and 2 Brothers, Sylvanus and Ammon.

She accepted Christ as her saviour in her youth, was a faithful member of the Fair Haven Amish- Mennonite Church.

Quickly and suddenly came the call
Her sudden death surprised us all
Dearer to memory than words can tell
The loss of a mother We loved so well.

Lester J. Miller

THE MILLER FAMILY

In our family circle, different occupations we do find, for God has given various talents to all mankind. As we exercise these talents, that God to us does give, we can be a blessing to each other as on this earth we live.

We have Ministers to instruct us out of God's Holy Word, his message always new, but still has oft been heard.

Some take God's word, to those who in Prison be, telling them of Jesus' love who came to set us free. Showing of their love to them in word and song, Explaining that our God forgives e'an though we have done wrong.

Some have Book Stores with Bibles and books to read just go Golden Rule Book Store and they will fill your need. We also have mechanics, that repair your mowers, tractors and such, to keep everything running smoothly, they really have a touch.

We have carpenters also, who can build you a house fit for a King, then do insulating and put siding on. They say "it's just the thing". There is also a Plumber, that your fixture will repair, such as busted pipes, or something else usually unaware.

Some are Electricians in that line, and they can find the trouble most every time.

Some are farmers tilling the soil, from morn till evening they labor and toil, but how thankful for the farmers we should be, for they are raising food for such that don't farm, you see. Some farmers raise eggs and some have cattle, for feeders, or milking, they build up your chattel.

Then too among these descendants some salesmen we do find, each selling products of a special kind, such as Vigortone, a mineral to be mixed into the cattle, horse and hog chow, it helps them grow and keeps them healthy somehow. Then another sells products of the "Erase Dirt" line, by doing your cleaning with these products, things will sparkle and shine.

And if we need an Auctioneer, he is available too, also a Realtor, who sells homes old and new, some are Managers or Foremen too of Firms and shops you see, Church Furniture, Cast Products and what else have we ?

We also have Truck Drivers that deliver Mobile Homes to other states, others haul products for Johnson Controls and what ever it relates.

Some do sweeping at the shops and get rid of the trash, while some others are self employed, bringing them more cash.

Now what about the women don't they work at all? O yes! The things we do are beyond recall. Cooking, baking, washing, ironing and sewing and mending too. Crocheting, knitting and quilting, there's also planting and weeding to do.

God has given all a talent, I've found this is true.

Sarah Mae Miller

GRANDMA'S CLOCK (CHRIS)

This clock belonged to My Mother (Grandma Ada) The story goes (as I gathered from My Brothers And Sisters). When My parents lived in Michigan

Near West Branch during WW1 circa 1914 - 1918, They lived in a drafty house in what Dad described As wilderness in a big dense woods with occasional Wild bears and other wild animals.

They would occasionally go into town for groceries And other supplies. One grocery store gave out tickets For a certain amount of groceries. When it was time for the drawing, the boys (Oba and Clarence) begged Mom to go to town that day.

When it was time for the drawing there was a Lady there with lots of tickets hanging over Her shoulder And She was sure the clock was Hers. When the lucky number was called, My Mom, a little Amish lady with one ticket had the right number, The Lady with many tickets threw them on the floor and stomped out of the store. The clock was a fixture in their house for ever after that.

I remember going to sleep many nights listing to the tic-toc and the chime counting off the hours and half hours sometimes late into the night when I couldn't sleep.

When ever it wouldn't run or needed cleaning and adjusting they would take it to "Watch Andy" (Oba –Fannies Dad) and He would get it going again. After Mom died We divided Her things and I ended up with the clock. I used to clean and adjust it Myself and it would run until granddaughter Rachel would run through the house and shake it then it would stop. It became too much of a chore to wind it up and maintain so I just admired it as it was for many years, at least it had the right time twice a day.

When I had My auction and sold My house, Lester and Leonard both wanted that clock and they bid it up way too high but Lester got it. When Mary had auction and sold Lester's things Leonard bought it, and I hope it stays in the Miller family for a long time.

Uncle Chris

MY CHILDHOOD MEMORIES (BARBARA)

The year of 1915, if I remember right, my folks moved to Michigan on the train and how we slept on the train. Grandpa went with us. Dad went on the box car with the animals and other belongings. Everything was so strange there.

We went to church in an open buggy, Uncle Oba's lived up there too, also Uncle Dave's and Uncle John's. I well remember the times, us and the cousins were together, those were enjoyable times. Several times Oba's and us went with the wagon and big team for a sight seeing trip,

We'd go through several woods and Oba. would hide behind the trees, and we children were afraid he would get lost. We nearly always went to West Branch to get supplies with the wagon, We'd take the oil drum along to get oil or kerosene. There were times Dad took us out to the lake in a boat to fish, we really enjoyed it.

We had to walk to school, back the lane through part of the woods, then through a field till we got to the road. It was about one-half mile to the one room school, and I was the only one in the Primer class.

One day us children were just alone at home, Oba. and Clarence had two pet chicks, they had made a harness for them out of strings and hitched them double and could drive them. They would unhitch them to feed them. That day when they unhitched them to feed them, a hawk came down and picked one up and flew away with it to he woods. Oba. ran after it , but he did not get the chick.

 We had a flowing well which had a pipe that the water ran all the time. One day when I was drinking, Clarence took hold of the pipe and as he jerked , the pipe cut above my eye, and Grandpa Millers were at our place at the time.

Grandpa sheared our sheep by hand in the spring, and the boys helped hold the heads down.

One time the young folks were at our place and had ice cream and two of the boys took one freezer of ice cream on top of the house to eat it.

I well remember when we went to school in Michigan. We had to go through the woods part ways, how the birds would sing, we were aware of the animals that were in the woods. I also remember how Uncle Oba. and us went to see some country west of us, going with two horses and wagon, drove through a thick woods that had wild animals which we saw along the way.

In the Winter Dad used to go to the lake helped cut ice to store for the Summer use. Dad often spoke of how dangerous work it was to be on ice with the horses.

When we lived in Michigan, Dad had rheumatism in the spring. Mother would go out in the field to harrow, days at a time, and Dad would get dinner ready.

We always thought it tasted so good, We wished he would get more meals.

Also remember when we moved back to Ind. Which was a long train ride. Dad went on the box car with the animals and other belongings.

Years later, in Ind. again Dad used to go thrashing. Dad used to run a thrashing machine for Sidney Zook for years. In the early mornings one of us children would take him, so he could make a fire in the steam engine and get things ready to thrash. It usually was quite in the mornings, birds were chirping and animals just waking up, what enjoyable experience it was. It was a different atmosphere than in the day time.

I remember yet when Dad had inflammatory rheumatism. How he could not move or talk then Mother would rest in a chair beside the bed during the night, then one night he could move again, Mother was so glad. There was also time when we had the measles, seven of us children and Mom were in bed at one time. Mom was quite sick . Cousin Noah Miller and his sister Mattie, helped Dad with the chores and taking care of the sick.

One time I was to chases the horses out of the yard, then one of them ran after me. Mom came with the mop stick to chase it away then the horse and I both stumbled at a small ditch. I fell one way and the horse fell the other way that was what saved me from being hurt.

Also as a young girl, I remember how Dad used to read Bible stories to us and read aloud out of the Bible. When Olen was a small boy he would run after Dad down the road when Dad went to town, then Mom would say "Barbara go get him" that was quite a chase, One time a neighbor caught Olen for me, I think that was the last time Olen went. A happy life in our younger years.

By Mrs. Barbara Hochstedler

ADA BURDINE YODER

My Mom and Dad, sisters and brothers were at Grandpa Miller's, Saturday night, when the next day, Sunday morning we were making breakfast when Grandma had a stroke. Then we found out Grandma had a light stroke Saturday night. She told Grandpa not to tell us that she had a stroke that night.

I also remember when I was a young girl and we knew Grandpa's were coming to our house, us children would watch the road until we saw them coming and Grandma would have this white hankie and wave at us as they were coming down the road. We were always so glad when Grandpa's came.

ADA BERDINE YODER (ADA)

My Mom and Dad, sisters and brothers were at Grandpa Miller's Saturday night, When the next day, Sunday morning we were making breakfast when Grandma had a stroke. Then we found out Grandma had a light stroke Saturday night. She told Grandpa not to tell us that she had a stroke that night.

I also remember when I was a young girl and we knew Grandpa Miller's were coming to our house, us children would watch the road until we saw them coming and Grandma would have this white hankie and wave at us as they were coming down the road. We were always so glad when Grandpas came.

CHILDHOOD MEMORIES (LAURA ELLEN)

I was asked to write something of my parents. There are many things a person could write. We had a full and happy life as children. There were also the sad times, but as small children, we seem to remember mostly the happy times.

I think what sticks in my mind most, was the time we lived in Howard Co. Dad had a blacksmith shop. The many hours we spent playing in the shop, or helping Dad, turning the handle on the forge, then putting the hot metal in cold water. (You younger ones don't even know what a forge is), Then there was watching Dad nail the horseshoes on the horses' feet.

Then after we moved to Millersburg, the time Mom was learning to drive a car. She went out the lane, turned on the road and kept right on turning in to the ditch. Also how things changed, remembering how Mom got her drivers license. We went to the Elkhart Police station, Dad went in and bought her license. The excitement it caused, when two baby sisters were born into our home, the sisters I had always wanted. But a year later, I married, then when the girls were 3 or 4 years old, one of them said to me one day, "Are you really our sister?"

When We told the folks about moving to Wisconsin, We could see it was hard for them to see Us go, but like the parents they were, they didn't interfere. But after we were here and they came to visit us, and Dad spent time fishing. He asked if our one building could be fixed up to live in. It had been a house once, but that time never came, because by the next summer, he was called away.

That last summer Dad was here, he and I spent three days out fishing together, and those days really stand out to me. We discussed many things those days, not knowing they were the last such time we'd have together. Going back to Indiana isn't the same anymore.

If you've had a Dad and Mom like mine you'd have loved them too.
Because they were gentle, tender and kind.
They were always there to help in times of need.
Knowing also for us they shed many a tear and time on their knees.
If you had a happy home like mine, you'd miss them too.
They are now gone from us, but the example of their lives and things taught us will go on for ever as we live these things, and we in turn, teach our children.
If we could only hold on to your hands once more, Dad and Mom;
What joy it would bring to our hearts,
but we know someday we may meet again,
if we keep true to the God they served.
And may we all meet together up there.

by Laura Ellen Gingerich

LESTER GUN (LESTER)

Hi,

Premier single shot 22 rife. Approximately in the mid 1940's. Brother Clarence Alma's parent's lived North West of Goshen on a farm. They had a good size truck patch (large garden) and my parent's went there to help pick up potatoes several times. Then there son William (Bill) Miller brought this rife out for us to target shoot. Well I went there lots of times and we went fishing and target shooting. One time I ask to buy it and he said no not at this time. Well several years later he was ready to sell and I brought it. A good light gun, good shot for it got me many squirrels and rats while I was at home and use it much. But something happen to the stock broke. I kept putting it together to make it steady and usable.

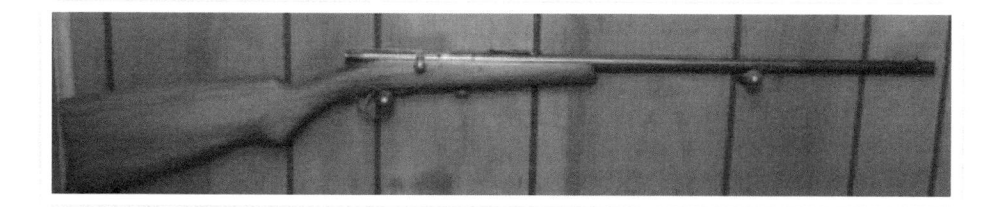

Then in 1946 Melvin and I worked at the some place in Goshen. Mel and I would run around together back then and Mel would always be asking questions about my sister Nancy. So in November (hunting time) I always would ask Melvin to go hunting with me but he wasn't interested then finally one day he said he'll go. So we went back to the woods behind my parent place south east of Millersburg. Mom and Dad lived on what was called the Brown place where this happen. We drove back to the woods instead of walking. We climbed a fence, Mel picks the gun up thru the fence and it went off. O boy we had to climb back over the fence again but Mel could hardly make it, finally I pulled him over and we crawled over to the car. Well I ran to the house and told Nancy about it. Nancy, Mel, my mom and I the four of us went to the hospital as fast as could thru stop signs and lights till we got there and help Mel walk in. The shot went into his left shoulder and landed between the mussels and blood vessels so no surgery was needed. The Goshen Hospital was on the east side of town on Second Street. I don't remember how long we were there but it was after midnight till we got back to the house. Thanks to God, Mel got thru this ok with no effects. Brother Chris took the firing pin out and I put it up in our living room on display. One day I showed and told this story with the history of the gun. Then his son Mike spoke up. He wants first chance to get that gun someday. Well I said not yet as it is part of my collection and memory of how the only time Mel and I went hunting. So one day I said, Mike you can have it. He said he wants it also for the memory of his dad. Then in 1947 Josephine and I got married the little later Nancy and Melvin got married. When I think back of 65 to 70 years ago how it was and what happened, well what do you say.

Happy time Mike

From Uncle Lester J Miller
04/18/08

OUR DAD C.J. (VELMA PECK)

Since I was C.J.s third daughter, I guess he thought having a tom-boy was better than having no boy at all. Anyway, I didn't mind I got to be with him a lot. C.J. especially enjoyed working in his big garden or garage after supper and I must have been his little tag-a-long ?

We'd talk and talk, I remember he answered some of my first questions about God and life in general. I still feel after years of schooling, that Dad's answers were the wisest and the best.

I felt so grown-up, when Dad needed me, to go along to give a paint estimate, My job was to hold one of his giant tape measures as he measured a barn or building and it was a real treat to go along with his big oil truck to make a delivery. I felt especially close to Dad when I was growing up, but I found out recently each of us five children feel this way. None of us was his favorite, he loved each of us in a special way.

Dad was sad when he couldn't afford some of the things we 'wanted' (not needed) as children. Now I realized all the money in the world couldn't buy the fun we had as a family. We rarely took vacations, but found time to go on Sunday drives, picnics, steam shows and other outings. We went fishing a lot, and C.J. must have really loved us considering he spent most of his time untangling our lines and recuing imbedded fish hooks !!

C.J. was very proud of his sons-in-law and daughters-in-law and especially his grand-children no visit was complete until he had held or teased them. When his youngest grand-child Linda was nine or so, our Kevin was born. C.J. used to find all sorts of excuses to stop by the apartment to see the baby. Dad died on the day Kevin was six months old, and my heart is sad that our boys never knew their Grandpa, and all he could have taught them.

We miss him so much, but we thank God we had our Daddy C.J. to give us the best instructions of all. A close knit family with a zeal for life, with love and concern for each other.

Velma

BREAKFAST NEWSLETTER   DEC. 11. 04
Tribute to Uncle Clarence (CJ) Leonard
Greetings of the Season;

The winter is in a lull and it be possible we may have a mild winter since, the past summer was cooler than normal ?

This letter is a tribute to Uncle Clarence (CJ).

Some of these items I may mention are to my best recollections and your opinions may be different and that is quite ok with me.

As a child growing up it was with joy we would go to "CJ's". My first Memories that I can recollect was them living in Waterford. He would desire to have others enjoy life as he enjoyed it with humor and good comeraderie.

I always like to hear his story about his car he had rigged up with a model T coil and as individuals would lean against his car he would give a jolt with it causing people to jump.

His love for the outdoors, desire to improvise with mechanical things, love for trains and steam engines, and the love for his family he would try to inspire others to the like. Should my memory serve me correctly he was into coon hunting for some time. His love for fishing inspired many to follow him.

I always have to think of the courage he had when his dissappointment at the Waterford homestead wasn't available to them. I believe he was misled in the understanding of this and that he was able to give his family a nice homestead in Jefferson area for them to enjoy and appreciate. It may be amazing how many individuals were inspired by him to collect old items. In the 1970's "CJ" asked me to come to come to "Logan Monument & Fuel" for 1year replacement for John Miller's absence. I believe it came to two years finally. One thing I learned about "CJ" was his compassion for individuals who were treated as "underdogs" and when he seen someone struggle with life's temptations he would not broadcast it or make an issue of it to the general public.

His faithfulness in serving "Jesus Christ" was in genuiness and humbleness. It was easy for him to converse with anyone. I was always amused in delivering fuel oil how he enter little children into a conversation and before we left we knew everything about the family.

Some people serve humanity in such a way that individuals are indebted to them in such a way that can never be repaid and I have found this for me in "CJ". He didn't laugh at me when I shared about my desire to go to auction school instead, set the profile for the first two auctions I did

My realization is this letter doesn't do justice for all he was and did. Just a glimpse of a very well-rounded individual that blessed many lives.

I'm sorry for missing the newsletter in November and breakfasts will restart in January. It seem like by the time the day was finished I was to tired or something like that.

Uncle Lester will be going to Ft. Wayne for surgery and Eugene will be going to Goshen for surgery on the 14th of December. I suggested to Eugene that they get walkie-talkies to keep each other updated. Let's remember them in prayer.

May the Christmas Season and the New Year Bless You.

Leonard and Laurie

UNCLE CLARENCE (ALVIN)

About the time uncle Clarence, Dad's brother got a car and went to the Town Line Church. He came to visit us a lot and even took us on a ride once. He also got a box camera and I and Laura stood in front of the car and He took our picture. He showed us how it worked. You look in this little window and see what you want to take a picture of, then push this little button. Then you have to take the film to a store where they develop the picture. We had that picture for a long time but somehow it got lost. Later I found it in Mom's picture album.

Uncle Clarence got a car with four doors and then took us and Grandpa's to Goshen, Once he took us and Grandpa's to Ohio near Charm. At that time a young man from there was courting dad's sister Mattie. His name was Elmer Miller. That family became good friends of our family. Elmer's sister was Sarah Mae, and had brothers John and Alvin. When Elmer and Dad's sister Mattie were married, Dad's brother Olen got acquainted with Sarah Mae. We all liked her and she was a good singer, also played the flat lap string harp, Also the harmonica same as Dad did.

When we went to Holmes county I had never seen so many big hills, There house was on a hillside where we walked in at the basement and way upstairs at a back door walked out on the ground. They had a squirrel cage beside the house and I was very excited to watch it run around and it had a wheel where it went round and round. Someone made a remark once asking how do you farm these hills? He said, we farm both sides of the field, a little joke we didn't forget. Another funny story that was heard around there, a farmer worked all day to get to the top of the hill but when he turned around to go down it only

MANY MILLER MEMORIES 169

took a minute.

On this trip we went to a place called the Dowdy (Doughty). It was a narrow valley with about six houses in it. A week before a flood had gone through the valley and damaged most houses. We went to see the damage and were astounded at what we saw. Only one house was habitable and we visited them. Their living area was upstairs. They described what they went through.

The power of the water had washed the iron bridge that was across the creek away down stream an left it sideway in the middle of the river. One house had the foundation washed away and the floor was just hanging on by one end.

We also visited a coal mine nearby, it was across the road from an Amish place they knew, So they took us inside. Walking down the tracks they used to push large boxes on rail wheels out to the place to the place where they dumped the coal and there the coal was sorted by different screens of different size holes it had. We saw the men digging and picking the coal loose and shoveling into the boxes. They teased me and put their finger on my cheek and said now you're a miner like them because they all had coal dirt on their faces. Then they took us to a building where the miners washed up. I had never seen so many black faces. They also changed their clothes because of the coal dust.

About that time Clarence began dating a girl whose family had just moved her from Alden, New York. She was Alma Miller whose folks were Henry and Amanda Miller. That was near Clarence Center about 25 miles east of Buffalo,

C.J.'s 1928 Chevy drawn by Velma Peck

N.Y., Henry was a good singer, we could always hear him above the everyone else. He was proud to say he once played for the New York Yankees. Henry's lived on County Road 19 about two miles north of Goshen, the house is made with rocks all around it. It's easy to notice. As it sits on a little hill on the west side of the road.

C.J. as we called him did not join the Amish Church. He was the only one of the family at that time that had a car so he often took us or Grandpa's places.

REMEMBRANCE OF YOUNGER DAYS (MATTIE)

While living in West Branch, Mich. Around the age of 3 or 4, one day while we were outside playing, we had bantams and small chickens and a chicken hawk came close and got one, we were so sad and despised the chicken hawk.

Another time when Mom and sister Barbara were planting beans in the garden, I don't know, did the boys help or not, but sure enough, here a rooster was following her and had picked them up almost as fast as she planted them.

On the trip moving back to Ind. Dad went with the train that the machinery and furniture was on, and Grandpa went with and us children on the passenger train. We had to wait a while at the depot, then a boy was walking around with a box saying," popcorn, and peanuts - popcorn and peanuts" So finally Grandpa got some, he was sitting with his back towards Mom. We ate some, after all I had to get a tummy ache.

In 1922 we lived North of Middlebury, on Mar. 7 that Spring. Uncle David's moved down from West Branch, Mich. They moved the first house south of us. The next morning we had a surprise, a little girl came to our hose during the night. It was named Nancy Alta, born Mar. 8, 1922. As Dave's Nancy, now Mrs. Seth Troyer was working for the folks.

On Sept, 7, 1930, on a Sunday morning, us children got ready, took the double buggy to go to Brother Obas for the day, we had a nice day, when we came home, there on the couch, was a little baby which we all admired. It did not have a name yet, but Brother Clarence seemed to look longer at the baby. All at once he said "it looks nice enough to be called Christie" Mom said that settles it as she had wanted a Chris before, but it never worked out that way. So it's Chris Jay, named after Moms Dad, Christian C. Miller.

Chris was 2 years old when we got married then we moved to Ohio, Grover Bowker taking us and our belongings. Chris didn't think it was very nice of him taking sister Mattie and Elmer to Ohio. About one half year later we were out at the folks, I was getting dishes out of Moms cupboard, and Chris was beside me when he saw a picture on a magazine in the cupboard then he said, look that's Grover, he's naughty taking you to Ohio. But Chris, guess you don't remember. But there are lots of good remembrances of family and a little fuss, which usually comes too. We thought Dad and Mom were so stern in instructing us, which I was never sorry. I appreciated my parents till death called them, and do so much more now since they are not here. I think we as parents were not stern enough, as it seems people are giving more lean now days as from what it was years ago. Our plea is, pray for our children and grandchildren.

Sister Barbara and I wanted to go to Goshen once for the first time. Mom let us go, so we got groceries, then went to the Dime store to look around, then a big husky man and wife came to talk with us and asked all kinds of questions, we had seen them in the street, but finally they left us. We decided that's the last time, or not right away, but really it was the last time. We went out the back door of the store and went home, afraid they would follow us.

Things can happen at times, which we do not realize until later years. In spring of 1932 Bro. Clarence wanted to go to Goshen, don't know why but I wanted to go along. Guess every thing was O.K. but anyway after we came home Clarence said, well Mattie, that's the last time you will go with me to Goshen. I just laughed at him. He said I'll write it down, and you know, it was on Dec. 22, 1932. Elmer and I were married. Then in winter of 1936 Clarence was at our place in Ohio. We were talking, joking and teasing then he made a remark if I still remember he told me that's the last time for me to go along to

Goshen, and he's going to write it down. I just laughed that he never did. Sure enough he reached in his shirt pocket and got a little note book that he notes and showed it to me, there it was "This is the last time Mattie will go with me to Goshen" Yes that's what made time go by.

EXCITING MOMENTS

Did your little boys ever wander away from you well ours did. At the age of about 3 or 4 years old. The rest of the children were in school, Elmer was in the field husking corn. The two boys Elmer Jr. And Marvin Dale and I were raking leaves. They started to jump in the leave pile and really had fun. Then I went into the house and told them to stay there and play or go with me, but it was so exciting to play in the leaves, we stay here. I said now don't go away, we won't Mom. I went in down to the cellar to get potatoes, till I got up I couldn't see my boys, so I checked in the barn and chicken house, back the drive which went to Long Lake. We lived Southeast of Lagrange. Elmer had taken the car down the road, parked it and walked back to the field, so I went to the car, no boys nor dog could be heard. I couldn't find the keys, so I just tooted the horn, thinking Elmer would hear it but he didn't. Here came a car with a man in it and he watched me so funny I thought, so I watched the car. Here it turned around and stopped, Do you need help he asked? I was so scared by then I didn't want to answer, so he asked me again, then he said, don't you know me Mrs. Miller? I'm Glen Bowman, your Production Credit Man, I recognized him then. I said yes, I can't find our two little boys. I think they went back to Elmer but I can't find the car keys to the car, and it's too far for me to walk, as I had ulcers or running sores on my leg, so I could hardly get around. Glen said get in my car, we'll drive back as far as we can to find them, so he stopped when he couldn't get much farther with the car, he got out so quick and said you stay sitting. I'll get the boys. He got out so quick. I didn't think about it until a minutes later what if they refuse to go with a stranger. I waited a while then got out of the car walked up the hill slowly till I could see down the hill and back to the woods. Soon Glen came walking one boy on each side.

When he came to the car he said, Mrs. Miller I got your boys but still didn't see Elmer, and he said that Elmer Jr. didn't want to come with him, but Marvin didn't hesitate, so Elmer Jr. Also went with him. When Elmer came home for dinner and found out of the news and scare, he said they would have had a distance to go yet , besides over a bridge across the river, that was not boarded tight, just enough planks for wagon or tractor to go over. The next summer when Perry went over the same bridge with the tractor he got too close to the edge of the planks and the tractor tipped on the side, but Perry thinking fast enough that he made a little jump and the tractor just missed him. We give God all honor and praise for saving our boys.

AN EXPERIENCE THAT HAPPENED ON JULY 24, 1973

I went to the Health Unit at Goshen, Ind. To get my birth certificate. She asked my name, birthdate, township, and parents name.

Well she got the Doctors Manual out and she found my birthdate, township and my parents name, also the Doctors name which was Doctor Teeters, but she said your name here is Minnie. Stunted, I said in ever heard of that name before. I was called Mattie ever since I remember. But after I got older, Mom had told me my name might be Magdalena, that's what it should be so she said we'll check in the school year books you probably started to school in year 1921. She found it but it was Magdalena, in year 1922 she found nothing, so went to 1923, there she found my birthdate. Then she asked did your father have a middle initial? I said yes it was Jacob J.C. and Ada C. You also have a sister Barbara and a brother Oba and Clarence, and I said yes. Well here your name is Mattie Miller, so she checked year 1924, there it was Mattie again, so she said she will have to fill out an affidavit to send to Indianapolis, or my name could be Magdalena there. I received my birth certificate on Aug. 13,1975 and with no trouble, and my name is Mattie** it just cost me $5.50 to get it done.

This write up furnished by Sister Mattie

** In her mother's Bible (in possession of Uncle Chris) Mother Ada wrote her name as, Magdalena J. Miller, was born Feb. 15, 1914 in the year of our lord,

REMEMBERING MATTIE ELLEN

My name is Bonnie Roderick. I worked with Mattie at the Hilltop (Restaurant) We worked side by side everyday. We became very good friends. She always said, "Bonnie you and Harry are family". We shared a lot with each other. We talked about our kids, our families, joys, and pains, almost everything. We often shared our relationship with God with each other.

Just a few weeks ago we were talking about heaven and she said, "I hope someday I will be able to get there". I said "Mattie, you CAN KNOW you will. Because of Jesus we can know we are saved and will go to heaven. I quoted John 3:16. FOR GOD SO LOVED THE WORLD, HE GAVE HIS ONLY BEGOTTEN SON, THAT WHOSOEVER BELIEVETH IN HIM SHOULD NOT PERISH BUT HAVE EVERLASTING LIFE. She looked at me as though she couldn't quite believe it was that simple. Mattie said, "I try to do what is right and HOPE that I can make it to heaven" I said, "Mattie We cannot earn our way to heaven" Eph. 2,8,9 says, "For it is by grace you have been saved Through faith. And this is not from yourselves, it is the gift of God-not by works, so that no one can boast" I said, "You believe in Jesus don't you?" She said, "Yes I do" I told her the plan of salvation is sometimes hard to fathom—We make it hard. Yes we need to live our lives as Christlike, but God knew we would fall short, so that is why he sent His son Jesus to be the ultimate sacrifice for our sin. If we accept him as our Savior, We will want to strive to live more like Christ. Asking for forgiveness and accepting Jesus is all we need to do.

I believe with all my heart that Mattie is in heaven with Bud. She missed him so. Mattie always said I believe it's what's in a person's heart that counts. She would do any thing for you. She was a giver, but it was so difficult for her to be a receiver. Mattie was stubborn, She was a very independent woman.

She loved her grandchildren, They were everything to her. She loved going to their games. She loved her kids and family. She worried about her Mom and Dad. She often said she wished she could move in with them and take care of them, But she knew her responsibilities were here. Mattie often said that she felt she was pulled in many different directions. To me it seemed she never felt the things she done were enough.

To her Brothers and Sisters and Nieces and nephews who helped who helped to take care of her Mom and Dad, You will never know how much this meant to Mattie, She spoke kindly of you often.

She loved Bud's family. Family was everything to Mattie, She loved the cousin breakfasts at the Essenhaus. You were of great support to Mattie. She lived her life the way she felt she needed to live it.

The church here was also very important to Mattie, The love and support you showed to her before and after Bud's passing were lifeline her, She told me she attended church as often as she could. I truly believe Mattie gave her life to God and is with Bud now.

MEMORIES OF CHILDHOOD DAYS (RUBY)

I don't remember everything in order, but we were at Grandpas quite a bit, I don't really know my age at some of the places.

Grandpas lived first farm south of South Haven Church, East of Millersburg, Ind. The barn was unpainted. I remember how they did the thrashing here, and I thought that thing never gets full the way they feed it. I remember sitting in the grain wagon, with Grandpa's old John Deere iron wheel tractor with spokes, He really liked to brag it up.

Before My Daddy died, the family got together and made apple butter in a big iron kettle between the house and shed , on a cold day.

Another time Grandpa came in, (it was cold outside) and he was gassed by exhaust fumes from a tractor running inside a shed. He could hardly breathe, (I thought he was going to die) I think this happened after our Daddy died and we lived there.

One time it had a lot of snow we went somewhere on a sled and horses, and took a lot of blankets to cover us, Once in winter, Grandpa and one of the boys came into the house with wild rabbits and teased us. In the summer, we took that big rug out of the living room and dragged it back and forth on the grass to make it clean. I remember helping Nancy and My Mom clean for a young folks singing there, and everything had to be spotless, and Grandma thought everything had to be done and not another singing soon.

I remember more family gatherings there at Christmas, one year, it had about two foot of snow you couldn't hardly go anywhere, also an ice storm.

 Their outdoor toilet was in the chicken pen and they had a mad rooster, so Grandma or Nancy had to go with us. This first place, grandpa had a buzz saw that he pulled with horses.

I remember the first time I went to Goshen with Grandpa and Grandma to the 5&10 cent store, they gave me cookies and candy to eat.

Grandpas lived close to the railroad, at night time I couldn't sleep good on account of the train noise. Later I remember more of Grandma here, in making meals and washing dishes, She thought She had it handy with a pitcher pump and a nice pantry. I also worked in the garden with Grandma. I was there one whole summer. Oba's Clarence was there at the same time I was. In the evenings we would all sit on the porch and swing, after supper until dark.

One time the whole family had a birthday surprise for Grandma, on a Sunday evening. She was really surprised and didn't feel She was worth it.

I worked there after Grandma had a stroke, Even though she was handicapped, She was so patient and kind and was glad for My help. There was a time when She didn't care if you helped Her. Grandpa did a lot of telling stories of long ago, he said they used to hide their money in the wood sheds or put in tin cans, then bury them in their gardens, they didn't trust banks, not sure when this happened. Also when I stayed in the winter of her stroke, I went with them to a lot of funerals of cousins and friends, I usually sat with Grandma and helped. People thought it was so nice I stayed with them. We also went to town, and Grandma went along at times. I did enjoy my stay with them. Grandpa died just after we had Arnold, so they never seen any of my children.

Sunday dinner about always was coffee and jelly bread, cookies, peaches and milk, this is what Grandpa wanted and Grandma agreed.

I went to young folks too at the time I was 16 years old, I used to sing a lot then, but kind a lost it over the years. Fri. is Our 54th wedding anniversary.

I enjoyed messing around in Grandpas shop before he had sale. I enjoyed my stay there, and I'm sure they tried to make it nice for me. Grandpa would carry the water for washing , also build fire to heat the water for me.

Grandpa had a lot of concern for his children even thou he was quite about it. He tried to make others happy, and we should do likewise. He was always willing to give a helping hand.

I could write more, but have to stop somewhere. Grandma loved Morning glories and flowers.

The things I remember of Uncle Oba, I always thought he was a lot like Grandpa in a lot of ways, a helping hand and jolly. When they lied in Middlebury, He and Fannie would come out to say hello and a few words as soon as we drove up with my folks.

As I think of Clarence, I always looked forward to go there, always felt warm welcome. When we were small he used to bark like a dog, behind a couch or chair. I first thought there was one. He was a lot like Grandpa, always had time to visit

Mrs. Simon (Ruby) Overholt of Auburn Ky.

Note: The cape worn by Ruby and Julia was among Mom's things and it is a square piece of brocade material that is folded in a triangle and was worn by Amish women in olden days like the present day capes that is now part of their formal dresses.

I REMEMBER

I first remember Grandpas living near the railroad tracks, South of South Haven Church E. Of Millersburg, Ind., We went there in horse and buggy from Millersburg, They had the thrashers that day. Grandpa cut his finger off. I was 5 or 6 years old, I think Grandpa was always good, and talked to us children gave us attention, but was firm.

I'll always cherish our Christmas gatherings when we got oranges and gum from them often, if not always, They gave us candy when we visited them.

I stayed with them one whole summer, when I was 12 years old or so. About the last year he was on the farm. He helped a neighbor with farming.

We hunted mushrooms, we hunted and hunted finally Grandpa came with a bagful (Of wind).

They were in town one day, and he called home, (the first time I answered a phone)He acted as a stranger. (Which was a long time joke). I know I enjoyed it there, and it helped me. Grandma was easy to work with and quite. I also enjoyed her short stay, living with us at home South of Topeka.

While thinking of Oba and Clarence, I remember them both as good Uncles, but never got close to them But they were happy and liked to tease and show slides'

Mrs. (Jerome) Julia Mast of Kentucky

Judy Beachy (Judy)

When we lived east of Emma Lake, I remember when Dawdy's came over one day, and after awhile Dawdy took us kids out to the garden, found a muskmelon, pulled out his pocket knife and we ate the whole thing, boy did that taste good! However Dawdy was in hot water with Mom and Mummy because it was before lunch.

One Sunday afternoon we were over at Dawdy's and us girls started to comb Dawdy's hair. Then we got some bobbi pins and had pin curls made all over his head, after we were done, in walked Preacher Dave Bontrager, everyone got a good laugh over it!

Dawdy would pick Theda and I up at times and take us home. Then at 13 &20 he would stop at the gas station and buy us each an ice cream bar, when we got close to where they lived, we would make us girls get on the floor of the car so Mary and Sara wouldn't see us, but it never worked, because we always managed to get to play together anyway.

Dawdy was a big tease, when Mike was born they stayed with us that day. Dawdy kept saying mom had a baby girl, not a boy.

I remember Dawdy's kitchen table, in the center was a tray with honey, salt, pepper, sugar, butter, it was always covered with a kitchen towel. Dawdy had a big white cup for his coffee. He would pour his coffee into his saucer and then back to his cup to help cool it down his coffee. He was always dunking something in his coffee also.

Mummy would at times say that the time sure flies. Always couldn't understand why she would say that, now I know.

Always enjoyed going out to Dawdy's shop, basically, just to see what he was doing/making that's where I discovered that he smoked Winston cigs.

Buzzsaw rig made by Jake and boys with Jonas, Chris and Betty

THE UNLUCKY SWITCHMAN (BY CHRIS)

When I was about 13 or 14 years old we lived on a farm with a long lane. The farm bordered the Wabash railroad on the north side, the track being only about a hundred yards from our house.

Many times when we were outside at night or early morning doing the chores we would wave our lanterns at the trainmen which always brought a response. Dad showed us how to swing a lantern we would hold it by the bail and swing it in a large vertical circle (the same as swinging a bucket of water without spilling any water) the engineer would always give us a toot toot and the brakeman would give an answer by swinging his lantern. We came quite acquainted with the trainmen and I think they looked for us.

Right directly to the north of our barn was the west switch of a mile long siding, where many times in the morning when we got up there was a west bound freight on the siding waiting for an east bound train, as soon as the east bound had passed a switchman would go and unlock and throw the switch and the west bound would proceed, now going west toward Millersburg first you went around a curve which went trough a woods and all the way was going up hill.

MANY MILLER MEMORIES 181

So when the train started out of the siding they had to get all the speed they could get to make it over the hill, but the engineer could not see when the caboose went through the switch.

Many times in the mornings we would watch the switchman as he would jump off the caboose, run to throw the switch, lock it and then run to catch the train which by that time was going pretty good speed, he would run as hard as he could and then grab the railing on the caboose and swing on. We would always say I bet won't make it this time. It was funny to watch but probably not for the switchman.

Well one morning when we were watching him running as hard as he could and the train gathering speed all the time he just could not catch it, he finally just stopped and put his hands on his hips and puffed as the train kept on going around the curve and out of sight, and we were wondering what he was going to do now, but he kept on walking.

Pretty soon tho we heard the train coming back to pick up their switchman and then they had to back up far enough to get a good run to make it back over the hill.

It was always fun to watch the trains and trainmen as they worked, they were always friendly and waved at us.

We didn't know that very soon the steam era would be over and we not see the living breathing steam engines working anymore.

Maybe not as efficient as diesel but much more interesting.

THIS AXE (CHRIS)

This Axe probably symbolizes the American pioneer and farmer more than anything else that they possessed. An axe was never very far away.

When the pioneers came to America they found a vast country covered by trees, then the only way to make farm country was to remove trees. And at that time about the only tool available was the axe. Using your imagination this was quite a formidable task. My parents told stories of My Great-Grandfather, Jacob Christner, clearing land here in Indiana by partially chopping / notching a bunch of trees and then falling them in a domino effect, This was a very dangerous situation. (His father was killed in such a situation in Canada) That axe head is still in Bro. Olen's family possessions.

This also required a sharp axe and with steel not tempered like is today the axe needed constant filing, that's why a lot of the old axes are only about half size anymore.

This axe was not only used for cutting trees it was also used for notching an fitting logs for cabins and other buildings. And later it was used for hewing timbers for barns and houses. A man with sharp axe could be a good craftsman.

Look in any old barn and you will probably see the marks of an axe made a long time ago. This axe was also used to blaze trails (mark trees with sign to show the trails and roads. They would also leave messages for others that would come later.

They also had to build bridges and fences to keep their livestock in and barricades to keep Indians and marauding animals out.

No wonder the first settlers carried axes in their belts and treated them with respect like that of a soldier toward His sword or side arms. Anywhere they went the axe went along and was always within reach.

An axe was enough to bring down a domestic animal like cow or ox, a sharp blow with blunt side of the axe would stun an animal and then they could kill it for butchering and the axe could severe bones and ribs during the butchering.

Most American farm boys (and girls) knew how to chop off the head of a chicken and they knew that would be good eating, fresh chicken and everything that went with it. Mmmm, Goood.

A lot of axes spent most of their time right outside the back door of the house, by the wood pile ready to cut firewood for the fireplace or woodstove and to make kindling for the cookstove. It was usually the job for young boys to keep the wood boxes filled to the top so they would last all night and some to start fires in the morning again.

This axe could also be very dangerous as there were many a man who lost a finger, thumb or worse. Some were by accident and some were plain foolishness, like the two boys playing, the one said lay your finger down here and I'll chop it off and when He did, He said to the other one why didn't you move Your finger and the unfortunate one said I didn't think you were going to do it.

My own experience was a close call, When I was helping Dad cut firewood and my axe was deflected for some reason and it cut neatly thru my overboot, my leather shoe, and wool sock but didn't have a mark on My foot!!!!

If you look closely at this axe you will see where it's scuffed deeply right behind the head, well that comes from "over reach" when a youngsters arms are growing it's very likely that the axe swings in a greater arc than expected and it ends on the other side of the wood, no amount of scolding seems to cure this problem. Then you will see wire wrapped around the handle at this point where it has weakened, from too many blows on the handle. Because times are hard and the price of a new handle cuts deeply into the budget.

As you look at the head you will notice that it's sharpened at an odd shape, that's because as you're chopping wood, when ever you miss or go thru the wood you're going to hit the ground and rocks etc. So the front part gets the most nicks hence-forth the more filing/sharpening.

The lowly axe is becoming a thing of the past, you hardly ever see one being used today with all the modern conveniences, like chainsaws, skilsaws etc. You probably wouldn't find them on anybody's Christmas list anymore.

But this axe was just as important to the taming and building of America as the Kentucky rifle, the covered wagon ,and the railroad.

I think it needs it's place of respect so the future generations can see what helped to make America great.

Uncle Chris

THE MAD DOG (BY CHRIS)

When we lived on the "Huffman place" that is a mile east of Millersburg and a quarter mile south, and I was about 10 or 11 years old, one day when my brother Lester was out plowing with our tractor, an International Harvester model 10/20.

He noticed a dog, sort of a big German Shepherd type, along the fence next to the woods-pasture just watching him as he drove by.

I guess Lester stopped and talked to him because the dog started following him around the field behind the plow.

Some time later Les got more friendly with him and the dog got on the tractor with Les and rode along sitting on the platform.

When Les came in for lunch the dog followed him up to the house and he was all excited about his new dog and wanted to keep him, but mom had other ideas SHE didn't want another dog around as we already had a little dog, a terrier named "Trixie", so mom told Les to take the dog out to the woods and chase him away.

Well that didn't seem to work because I don't think Les tried too hard and the dog thought he found a new home, I think Les was also feeding him a little.

This went on several days and mom was getting irritated she apparently did not like the dog at all and she insisted that dad get rid of him.

When the dog realized he was not welcome he always stayed with the tractor, because of steel wheels and very slow speed we left the tractor out in the field we were working, especially if it was some distance from the house.

Well this day after our noon meal and feeding and resting the horses we got ready to go out the field again which was the farthest field from our house.

Well Dad took along his old shotgun and two shells. when we came out to the tractor the dog was laying by the tractor so Les and I tried to chase him away. I remember we would tell him to run quick or dad would shoot him. but he only went a little way then sat down and looked at us confused. then dad raised his shotgun thinking that he could scare him away.

The load of shot hit just below him and I don't know if any hit him or not but this seemed to infuriate him and he stood up turned around a few times then came right at me as I was the closest one to him.

He was growling with his teeth bared and snarling, just as he was making a jump at me dad had reloaded and fired hitting the dog in the throat. I still remember seeing the charge hit and seeing the blood fly from his throat.

He sat back on his haunches then turned and charged at Dad, well he had no more shells so he turned the gun around and began swinging it at the dog like a club, The dog kept charging again and again first the fore arm of the gun flew off then finally stock broke off but the dog still kept coming. I remember seeing the dog trying to get up the last times still snarling and growling. The last time he raised up then fell back down he was probably dead by then by loss of blood because he was bleeding profusely all this time. I believe if he had not bled to death he would have caused us serious injuries.

Dad stood there awhile and asked if Les and I were alright I think he was shaking aplenty right then. After making sure the dog was dead Dad told us to take him back to the creek bank in a gully, ever afterwards when we went past the place we would very cautiously look over the edge to make sure he was still there. I don't know about Les but I was very frightened when this was going on. and I thought Dad was very brave and courageous but then I guess he was protecting his two boys.

The shotgun could not be fixed to work properly afterwards so dad traded it for another one and that's another story

PERRY'S SHOTGUN

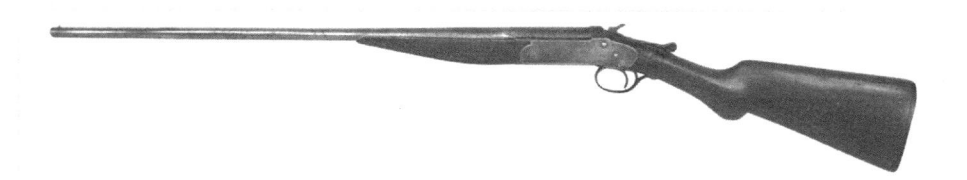

I always called it Perry's shotgun because of the circumstances and the things that happened because of it.

It all began when my brother-in-law Perry Yoder was married to my sister Alice I don't know when or where he bought this Iver Johnson "Champion" 20 gauge single shot. but it was a nice light weight easy handling shotgun.

It was not long after when he developed an Aneurism in his head and died soon afterwards.

So when Alice had sale to sell Perry's things Dad traded with Alice for his shotgun, Dad immediately liked it because it seemed he was more ready to go rabbit hunting when he got the chance. He carried it on the John Deere "D" tractor where it got a lot of scratches. and one time he shot a red fox from the tractor as I was driving the tractor the fox was surely not overcome by speed !

When I was a teenager my brother Les and I used it many times on our forays to the hunting grounds and brought back many rabbits for our dinner table.

When dad passed away and Mom had sale to sell some of his things, I determined that I wanted that shotgun and I bought it even though I didn't have much extra money at that time.

As time went by I didn't really use it as much I would have liked to. Then in 1970 we bought a van camper and went traveling throughout the western states. and I took the gun along in a case in the back corner and didn't think too much about it.

One time I think about '72 we decided to take a trip up to Ontario, Canada to do some fishing in some of the old spots we used to. When we entered a Provincial park that we wanted to go fishing they had a sign saying no guns allowed in park but in the U.S. you can take a gun most anywhere if it's cased. I didn't know what to do but decided to chance it, sure enough before we got to the camping area a park ranger stopped us and asked if we had a gun and no amount of pleading and reasoning but he confiscated my gun! I was just about sick but he told us who we could talk to the next day in Chapleau. Ont. but when I talked to the man, he had no good news either even though I paid my fine I still could not have my gun, he said I would have to wait until they have sale to auction off confiscated guns then he would see if he could purchase it and send it to me.

Well, it was several months later that I finally heard from them, they said they would send it to me. when it finally arrived I had to go to the Elkhart UPS office and pay the auction price plus import duty and ups shipping I don't remember how much it was but I probably could have bought several good guns for the money I spent. In Canada the laws are different and they are not always not very tolerant of "Staters".

A year or so later bro. Les and I were deer hunting in the Pigeon River Game area. in Indiana you use a shotgun with slugs and it doesn't matter what gauge you use the results are about the same.

As I was stalking through the woods I spooked up several deer and I could see one of them had antlers so I aimed and fired, the deer just seemed to disappear but when I came up to him he was down so I gave him another shot just to make sure he was dead. I started to dress him out when Les came to help me and drag him out to the road and my van.

When we dressed him we discovered he had a broken leg some time ago and had all healed over and had a knob like on his elbow he also had an abnormal antler.

When I went to check him in at the ranger station, the ranger asked if I wanted to reject him and try for another one

I asked if he thought there was something wrong with the meat and he said no he thought he was in good shape considering but he said the deer would probably have not made it through the winter.

I had no trouble with that as I was just as happy with this as I was looking forward some "deer burgers". we took him over to Bill Nisley and he butchered him for us and we enjoyed the good venison!

I took the antlers to Wayne Andrews and he mounted them on a board and I have them hanging on my living room wall along with a photograph of the event.

I have spent many hours in the woods and field with the Iver-Johnson chucked under my arm and enjoyed every minute, but it will still always be Perry's gun.

ADDINUM;

Niece Betty Yoder/Schrock always said she wanted that gun as it should stay in the family and it would be a remembrance of Her Dad.

So when I down sized My possessions, I sold it to Betty and She was happy and hopefully it will stay in the family along with the story that goes along with it.

THE SURPIRSE BLIZZARD (ALVIN) DECEMBER 1929

It was a beautiful mild morning for December. There wasn't a cloud in the sky . So far this winter there had not been much cold weather that lasted more then a week.

So Oba and his brother in law, Simeon Miller went to Goshen on the Shoup bus that ran on a regular schedule between Topeka, Shipshewana, Middlebury and Goshen. This was convenient for the Amish to get to town and return the same day. They arrived in Goshen at the Olympia Candy Kitchen around nine o' clock and strolled down through main street which at that time had all the main stores.

Soon after they arrived it got dark and began to snow. not very heavy at first but soon it was real heavy with large flakes. then about noon Oba decided he needs to get home before it gets too heavy and maybe couldn't get home so he saw Simeon and said we should get home as my wife's due with a baby any time now. and Simeon said he was going to stay in town with their aunt who lives on north 2nd. street right next to the railroad . Oba made his way to the Olympia Candy Kitchen wondering if the bus might be there. It was and Mr. Shoup was there in the restaurant . Oba said he was worried that the snow was getting deep and think we should go home before we can't get out of town.

Mr. Shoup agreed and Oba said, I need to get home as my wife Fannie is due with a baby any time now. Mr. Shoup said, so is my wife. maybe we should leave even before my scheduled time to go. So they left with only Oba as a passenger, going out east Lincoln street which became route 4. the snow was quite heavy wet snow, which is harder to cut through. The farther they went it slowed the bus and finally stalled at Yoder corner which now is county road 35.

Then Oba decided to walk to his parents place and borrow a horse to ride home. he had to walk a mile south and two and half miles east to his parents farm which was Jacob J.C. Miller. this is also known as Fish Lake road. He rode a horse without a saddle to his house which was on the Cableline road then now is 100 s. and two miles east of the county line of Elkhart and Lagrange counties . he arrived late but glad to be home and so was his wife Fannie.

It snowed hard all night and by morning cleared up but everybody was snow bound . then his wife began to have labor pains and he called Doctor Peters in Middlebury to come using the neighbor, Early Bontrager's phone. . He said there was no way to get there that the roads were all snowed shut. but then he said he thinks the road is open to Shipshewana and then going south should be better so he's going to try and make it that far. that now is state road 5.

Oba and a neighbor then went out on their road and shoveled through the deep snow from their house to the road coming south of Shipshewana. that was almost two miles . the doctor arrived just in time and the baby boy was born which they named Alvin.

This story was so often repeated that I have memorized it to this day. often when people asked when I was born, I'd say, do you remember the blizzard of '29 Christmas ? they'd say yes , well that 's when I came into this world

Written by Al Miller, Greencroft

IT HAPPENED AT JAKE'S (CHRIS) A TRUE STORY

Related by one of the witnesses who has a not too sharp a memory but a good imagination.

It was a nice warm sunny day to have a Jacob J.C. Miller family reunion at Jake and Barbara's farm in Howard County Indiana, in the summer of 1967. Quite a few of the families traveled from up north (Elkhart County) some carpooled and some of the youngsters even came by motorcycle. After getting reacquainted and catching up on the latest news, everyone gather around the dinner tables and enjoyed a delicious meal, as usual, potluck and we didn't seem short of anything. After dinner as the ladies cleaned the dishes, the youngsters went out to see Olen Hochstedler's trained pigeons. Olen had made a ramp onto his little wagon, like a cattle shute, and the pigeons walked like they were going to market.

Some of the men decided it was time for some serious baseball in the cowpasture, so off they went using dry cowpatties (I think) for bases. Clarence the ace pitcher and others thought to be semi-professionals kept the lively game going with a lot of bantering and hee-hawing.

After that the two oldest brothers decided to have bicycle race, my guess is that Clarence told Oba he could beat him out the lane and back with his hands behind his back, and Oba reminded him he was the older brother and he was going to show him how it was done.

So with Julia as official referee and score keeper and a good sized crowd of spectators, off they went. Soon Clarence got into trouble either hitting loose gravel or pantleg in the chain, he ran off the side of the road and in to the fence, and thinking he was severely injured by his moaning and groaning, we rushed out and helped up to the house where we laid him on a table and examined him head to foot by local quacks and family professionals, (there were no medical facilities nearby.)

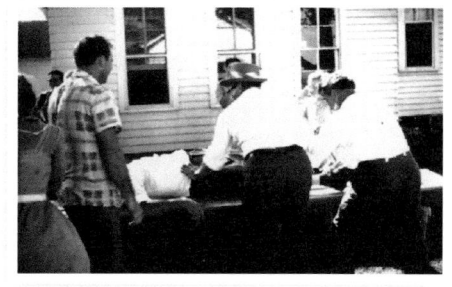

It looked pretty bad for a while, but it turned out not too serious after all.

After much groaning and moaning and lots of sympathy from everyone present, everyone agreed he would live to bike another day and then they decided some of Jake and Barbara's delicious afternoon refreshments was in order.

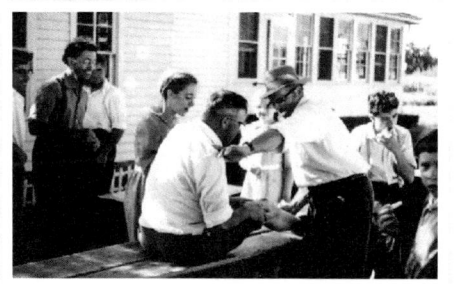

Towards evening we all got in our vehicles and headed back north to our homes with some more good memories of another good reunion at Jake and Barbara's farm.

Uncle Chris

OLEN'S STORY (AS TOLD BY OBA'S ALVIN)

One time I got to sit down and had a nice visit with Olen. And he told me a lot about himself that probably no one else knew or remembers. As you remember he was born May 30th, 1918 while grampa's lived in Michigan near West Branch. He was the second brother of my dad's after brother Clarence. Then they moved to Lagrange County, Indiana.

When he was fourteen years old his sister Mattie married Elmer K. Miller from Sugar Creek, Ohio. That was December 22, 1932. Grandpa's and Ben D. Miller's had to leave that afternoon to attend the funeral of grandma's stepmother in Ohio. The young folks all went along. That evening after supper the young folks usually gather to sing hymns. Though Elmer and Mattie the folks knew that Elmer's sister Sarah Mae and her brother Alvin sang a lot. So they got them to sing for them.

Once when they had just finished a song, Olen popped up and said, "oh you think your smart", it dutch. He didn't realize how it sounded or how much it hurt them at the time. Then through the years and with Elmer's they were it touch at times and four years later began to write to each other. That started a courtship and almost three years later they were married. They were married December 12th, 1939 and moved to Indiana. Soon after Elmer's folks and Alvin moved from Sugar Creek to Topeka, Indiana.

Olen likes to tell about some interesting things that happened when he was a small boy. When he was about five years old he went with his big brothers Oba and Clarence to haul gravel from Maude Webbs pit on county road 37, and they let him drive the team of horses which was a big thrill for him.

One day at the time they lived on Noah Bender's place on State Hiway 13 North of Millersburg that the folks had gone to help butchering at uncle Dave's, Oba wanted to solder a gasoline tank in the shop, I was in the shop with him. It exploded and blew the bottom of the tank out clear to the barn at least fifty feet away. And the shop was all in flames and then suddenly it was over. I didn't go to school yet and Oba was about 16 years old then.

Another thing I remember when I was in the second grade one Sunday, Oba and Clarence and Sandy Yoder with a few other boys at our folks were telling how they carved their names on a school desk when they went to school. So I decided if they have done that then I could do it too, this was at the Brown School on St. Rd 13.

At this time Oba was 18 and Clarence was 17 years old, I had a new desk and guess I had a new pocket knife well, anyway I carved my initials O.J.M. on top of my new desk. I also bored a hole in it so I could stick my pen down into my ink bottle. This was a common thing for boys back then. I'd try to put a book in front of me pretending I was reading while I was cutting and carving the hole.

My teacher was Vivian Johns (now Mrs. Ezra Slabach) did not notice until a week later. Then she wondered why I did this and asked me, I said my brothers did it when they were in school. Well she didn't punish me but said " you'll have to keep this desk as long as I'm in her room".

But anyways as things happen, it was in 1975, I'd say, Leonard (his oldest son, now an auctioneer) and Kenneth Stutzman had a donation sale at Lloyd Troyer's and someone brought some old school desks that had been stored, were real old looking and their finish worn off. It so happened that Leonard bought some of them and a neighbor agreed to haul them home for him. I was there and was asked to go along and help load them. Then while loading I took a close look and saw some markings on it and one had my brother's initials on it. Then while unloading at home I took a closer look at one of the little ones and it had my carvings on it. So I bought these two and my wife Sarah Mae refinished them. Now we have them for keepsakes.

Through the years he worked at Shasta Travel Trailers in Goshen as a saw man. He once worked at Nees and Hickerson which made struss rafters as a saw man. He was so nervous that the owner, Mr. Hickerson said he's afraid he's going to get hurt and asked him to leave which he did with out hard feelings. I helped with him as I (Alvin) was the foreman then. Olen apologized because he didn't feel too comfortable here and went back to Shasta.

Sarah Mae has written about 1,000 gospel songs and played the harmonica and the old type stringed Autoharp, she got some songs published, and had a book which we still have. She still sang and yodeled until she was too old. they attended Fairhaven Mennonite church. Olen died September 1st, 1993 and Sarah Mae died June 5th, 2011

Written by Alvin J. Miller

ROP THE HORSE STORIES (CHRIS)

ROP SAYS THAT'S ENOUGH

Dad had a sort of all-purpose horse called "Rop", he could be used as a work horse with-in a team of other horses to plow or harrow etc. and then could also be used as a buggy horse. Well Olen used him and a single buggy during his courting years and as boys did in them days, he liked to find out who had the fastest horse, and Rop did not let him down. so therefore he was used to passing other buggies and taking the lead.

One time there was a funeral, I think one of my cousins, anyway going to the cemetery, Rop could not understand why he had to stay in line that long. After the burial at Clinton Union Cemetery, on the way home, and I remember riding with Mom and Dad a few buggies ahead. When we heard an awful commotion behind us and there was Rop kicking like a mule and would not stop until he had the front of the buggy and the shaft all kicked to pieces, by that time he was thru the ditch and out by the fence tangled in the harness and totally exhausted.

I remember we had to pick up the pieces and pull the buggy home where dad and Olen fixed it up again, Dad told Olen with a sly grin, maybe you shouldn't be racing with him so much then he wouldn't do that. But Dad loved a good horse race as much as anyone. Olen said at that time if Rop wasn't worked during the week and ran on weekends he would be ornery come Monday and it didn't seem to bother him.

We had Rop for I don't know how many years and when he became too old to work Dad put him out to pasture and would not sell him. Dad said Rop worked hard for me all his life, now I'll take care of him until he dies.

OLEN'S COURTING DAYS AND ROP

When Olen was courting Sarah Mae. She worked for her brother Elmer K. and Mattie, they lived just west of Topeka and we lived SR. 13 north of Millersburg, 'about ten miles'. Many times when Olen was on the way home in the middle of the night, in his single seat open buggy with one kerosene lantern !!! He would start out, then tie the lines to the whip socket and lat down on the seat and go to sleep. Most times when Rop came home, after crossing St. Rd. 5 and a short distance on St. Rd. 13, and stopped suddenly Olen would wake up and find himself by our barn.

But one time Rop didn't go to our barn but went to the barn of his former owner. This time it didn't wake Olen and when those people came out to do their chores they found Rop standing patiently by the barn and Olen sound asleep in the buggy. Olen made a hasty dash for home and had some explaining to do before breakfast.

THE BUGGY RIDE

This is a story I hear many times as a boy growing up. It's about my brother Olen when he was a youngster, it seems he wanted to go along with dad to visit uncle Levi's and Dad wanted him to stay at home. So Olen decided to lay down in the buggy box under the seat so Dad wouldn't know he was going along until they got to Levi's place.

But Olen fell asleep and didn't wake up until they got back home. Dad never realized Olen was along and Olen didn't find out he was at uncle Levi's until he got home.

SARAH MAE

Before Olen and Sarah Mae were married, I remember, one Sunday after church when we came home. Olen was giving Sarah Mae a ride on our little coaster wagon and having a jolly good time. When Lester standing away from them had a baseball and was going to throw it at Olen (I guess) and when he threw it, the ball hit Sarah Mae in the left temple, she immediately collapsed and was knocked out cold. Well Olen and Lester were surprised and very scared boys.

Olen picked her up and carried her into the house and applied first-aid ie., cold wash rag etc. and shortly she came around again, much to the relief of Olen and Lester. Especially because they thought surely that Lester had killed her. But everything worked out just fine after all, and everyone was forgiven. And soon after that Olen and Sarah Mae were married.

THE STRANGE BUNDLE

When I was just about 4 or 5 years old, we lived up by Sand Hill school in Lagrange County. On a farm with a white barn and a green house if I remember right. One evening my brother Olen and one of his friends came to our house after supper and they had a big bundle with them. They brought it into the living room and laid it down and said they had something for me if I could guess what it was.

It was wrapped in a large blanket about 18 inches across and about 4 feet long, it would sometimes move real slowly, rolling over and sometimes would jerk in a funny way, then it would make funny noises sounding like cat meowing and then it would growl like dog and then make other strange noises. I kept guessing what it might be. A cat or a puppy perhaps ? no they said keep guessing. As it rolled it came open just a little at a time but I could not guess what it was. Finally as it opened a little more a hand appeared, then they wanted me to guess who it was. I guessed my friend Robert or Alvin ? no they said keep guessing, as a shoe appeared from the blanket .

But I could not guess who it might be, so he finally crawled out from under the blanket and stood up. Here it was a neighbor boy from where we used to live in Clinton Twp. By the name of Omar Miller, I was so surprised I didn't know what to do. So we played for a while that evening because he had to go home early the next morning with my brother Olen.

MEMORIES OF DAYS GONE BY (OLEN)

The time the folks lived in Michigan.(This I just know from hearing Dad talk about it in later years). Dad was buzzing wood and thrashing beans. This was in the years 1917 and 1918, There were not too many Amish people around there, most were Catholics.

The main crops raised were potatoes and soup beans as the growing season was short and they just had one railroad and means to get their produce to the market was poorly. Often the people were not able to pay him for his work for a long time.

I remember of hearing him say how they took their cream to West Branch, then from there it was shipped to Lansing, then they had to wait until they got their cream check from Lansing. I remember of hearing him saying how one woman paid him seven dollars all in pennies for buzzing one day.

Due to poor market for their crops and the short growing season, the folks moved back to Indiana again. Uncle Oba's, Dave's and Johns lived up there too, but they also moved back. After the folks lived in Indiana, Father done a lot of thrashing in the summer time. He operated a thrashing machine for Sidney Zook nearly all his married years. During the Winter months he would be buzzing wood, he was known far around for his work.

I believe Dad was buzzing wood for other people most of his married life when health permitted. In fact the morning he passed away, he was getting ready to go to Jake Marner's to buzz wood. I remember when we lived on Noah Benders place, how Dad would hitch old Sadie to the buggy early in the morning and have me go along to where he had the thrashing rig, so he could fire up the steam engine, when enough steam was built up, Dad would let me blow the little whistle on top of the boiler, then I was supposed to start home again with Sadie. I never was allowed to stay to blow the big whistle as Dad always done at seven o'clock. They changed over to an Oil-Pull tractor soon after the safety plug blew out of the steam engine.

Dad would often stay over night while he was away thrashing.

By Olen

Epilogue

What will the future hold, don't we all like to know? What will our future generations be like, will they up hold the principals that our ancestors lived by, will they know how and why our ancestors suffered and died, will they be willing to stand firm in their faith and suffer the consequences. Will the churches be able to hold together their flocks or will they scatter to the winds and whims?

Will the families be able to hold together against storms of strife, as every parent knows and worries after their children as they leave the nest, will the next generation be able to stand against the forces of evil?

The old Devil, Satan, Lucifer is still as active as ever, watching whoever he can waylay and trap in his clutches.

As in times past as I read the stories of the generations of people before me, I see that most stood steadfast, but there was always some that strayed from the narrow way and succumbed and wandered away, I can see why God worries about every one of his sheep and morns when he doesn't know where they are. There is hope that as they were brought up and taught as youngsters they will remember the faith of their parents and ancestors who prayed for them before they were born.

So with the churches to day while some cling to the teachings and sing the songs that the Martyrs sang and separate them selves as much as they can from the world, there are also some that conform to the wishes and whims of the world not wanting to offend anyone and follow a self centered society. Blending in and hardly noticed among the neighbors, without a real identity as believers.

As always God is faithful to those who believe and are faithful and we need to know as the Anabaptists did in their time that to be a Christian means that we are willing to suffer as Christ did and willing to give our life for our faith, and faith is being able to hold on even when the flames come about your feet, because the Devil is still right there to catch you at the very last minute.

But I do have faith in the generations to come as I see the younger generation come together and clinging to the older generation and wanting to draw the strength of faith from their forefathers and planning for the future of their offspring.

God has been faithful to his people throughout the Bible times and as well as since then and I see his hand in protecting and guiding his people no matter where they are both, now and forever, Amen.

Uncle Chris